The New Age of Space Exploration

Azhar ul Haque Sario

AF706209

Copyright

Copyright © 2025 by Azhar ul Haque Sario

All rights reserved. No part of this book may be reproduced in any manner whatsoever without written permission except in the case of brief quotations embodied in critical articles and reviews.
First Printing, 2025

Azhar.sario@hotmail.co.uk

ORCID: https://orcid.org/0009-0004-8629-830X
Disclaimer: This book is free from AI use. The cover was designed in Microsoft Publisher

Copyright Disclaimer: This book is independently produced by the author and is not affiliated with any space agency, corporation, or board mentioned herein. All references to companies, missions, or data are made under nominative fair use for educational and informational purposes.

Contents

The Dawn of Commercial Spaceflight

The New Argonauts: Sketching Star Maps in the Age of Private Skies

Listen closely. That thunder rolling across the sky? It's not just nations planting flags anymore. It's the sound of blueprints turning into launchpads, of boardroom dreams taking fiery flight. We've entered a thrilling new chapter in our cosmic story, penned not by superpowers alone, but by a bold crew of entrepreneurs, engineers, and innovators – the private space pioneers. Forget the old, state-drawn maps; these New Argonauts are charting their own courses, driven by a potent cocktail of vision, guts, and the audacious belief that 'up there' is the next 'here'.

Leading this charge, often with the roar of a returning rocket, is SpaceX. Think less 'company,' more 'force of nature' harnessed by Elon Musk's near-mythical ambition: making us neighbours with Mars. They didn't just build rockets; they shattered the old rules. Watching a Falcon 9 descend, a pillar of fire gracefully kissing the Earth it just departed, wasn't just cool tech – it was the sound barrier of cost being broken. Suddenly, space wasn't just for the chosen few. And they don't rest. Look up: Starlink weaves its celestial tapestry, aiming to beam the internet to the most remote corners of our globe, already connecting hundreds of thousands (as of late 2024). Look ahead: Starship, that gleaming behemoth, endures spectacular, data-rich ballets of fire and flight, each test forging the path, step by explosive step, towards lunar bases and Martian horizons. SpaceX isn't just launching things; they're laying the railway tracks for humanity's multi-planetary future.

Sailing a parallel, yet deliberately calmer, current is Jeff Bezos' Blue Origin. Their mantra, "Gradatim Ferociter" – step by step, ferociously – speaks volumes. They too have mastered the art of the reusable rocket with their elegant New Shepard, but their initial focus felt more intimate. They opened the door to space, not just for satellites, but for people. Imagine William Shatner, Captain Kirk himself, gazing back at Earth, fulfilling a lifelong dream – Blue Origin made that happen, offering precious minutes of weightlessness and a perspective-shifting view. Their recent flights hone this experience, making the journey smoother, the turnaround faster. But don't mistake patience for lack of ambition. Their powerhouse BE-4 engines now give roar to ULA's new Vulcan rocket, proving their engineering muscle feeds the entire ecosystem. And while Mars is SpaceX's headline, Blue Origin's gaze is fixed firmly on the Moon, with their planned New Glenn heavy-lifter and Blue Moon lander aiming to establish a permanent, bustling presence on our nearest celestial companion. Their vision? Millions living and working off-world, starting with our silvery neighbour.

But this isn't just a tale of two titans. A whole constellation of innovators is now sparkling. Rocket Lab, under Peter Beck's leadership, brilliantly saw the need for agility. They became the maestros of the smallsat symphony, offering dedicated rides to orbit from their launchpads scattered like stepping stones across the globe (New Zealand and the US). Their Electron rocket proved that mighty things come in smaller packages, even plucking boosters from the sea for reuse. Now, they're stretching their wings, developing the larger Neutron, refusing to be typecast.

Then there's the dream of Sierra Space's Dream Chaser, an elegant spaceplane gliding home, and Axiom Space, literally building the next generation of orbital real estate. These companies, and countless others crafting orbital tugboats or dreaming of micro-gravity factories, are weaving a richer, more resilient cosmic tapestry.

Taking this new space age truly global isn't easy. It's a voyage demanding more than just powerful engines:

The Relentless Quest for Cheaper Tickets: Innovation can't sleep. Reusability, smarter factories (think 3D-printed rocket parts!), and slicker operations are the wind in their sails, making the dream of space more accessible with every launch.

More Than Just a Launch: Relying solely on rocket rides is risky. The smart money is diversifying – whispering internet from orbit (Starlink), offering zero-G getaways (Blue Origin), watching over our planet, building things in space, maybe even mining asteroids one day.

Sailing Together: No captain sails entirely alone. Partnering with NASA, ESA, and other established players means sharing charts, pooling resources, and navigating the tricky currents of international space law.

Earthly Rules for Heavenly Ambitions: Every nation has its own rulebook for the skies. Going global means mastering a complex symphony of regulations, launch licenses, and broadcast frequencies.

Forging the Lifelines: As launches become routine and missions more daring, you need reliable supply lines – the unseen arteries feeding these metal beasts with parts and fuel, stretching across the globe.

The Brightest Minds: This isn't built with brute force alone. It requires attracting a global brain trust – the sharpest engineers, coders, and mission controllers from every corner of Earth to fuel the fires of innovation.

These pioneers, these New Argonauts, aren't just flinging metal into the void. They're launching possibilities. From SpaceX reaching for Mars, to Blue Origin settling the Moon, to Rocket Lab giving tiny satellites their own ride to the stars, they are fundamentally rewriting our cosmic address. Their recent triumphs aren't just headlines; they are waypoints on a journey proving that the final frontier is opening – for business, for discovery, and maybe, just maybe, for calling it home. Scaling this dream across our world requires a blend of engineer's precision, entrepreneur's hustle, and diplomat's grace. The countdown hasn't ended; in many ways, it's just begun.

From Throwaway Giants to Spacefaring Phoenixes: The Art of Bringing Rockets Home

Imagine building the most incredible, powerful car imaginable. A masterpiece of engineering. You drive it once, floor it down the highway in a blaze of glory... and then steer it straight into a crusher. For decades, that's pretty much how we treated space travel. We built magnificent metal titans, filled them with the stuff of controlled explosions, and cheered as they tore a hole in the sky, only to accept that most of this exquisite, expensive hardware was destined for a fiery funeral in the atmosphere or a silent grave beneath the waves. It felt... necessary. But profoundly wasteful.

Then came the whisper, the audacious thought: What if they could come back? What if these sky-piercing giants could learn to return, not as debris, but as vessels ready for another voyage? That question didn't just tweak the engineering; it ignited a revolution. It's changing everything about how we reach for the stars.

Think about it: watching a skyscraper-sized rocket booster, moments after blasting towards orbit at mind-bending speeds, choose to come home. Seeing it pirouette in the thin upper air, then plummet back towards Earth, only to fire its engines in a final, defiant roar against gravity and touch down, gently, on legs barely wider than its own base. Sometimes on solid ground, sometimes on a tiny, autonomous barge dancing on the ocean waves. It looks like magic. It feels like magic. But it's the breathtaking art of engineering pushed to its very limits.

How does this metal beast tame its fall? It starts with a cosmic U-turn, using tiny puffs of gas to flip itself around, aiming its engines back towards the ground it just left. Then, as the air thickens, intricate 'grid fins' – strange, lattice-like wings near the top – unfold. They aren't just steering; they're biting into the supersonic wind, tasting the atmosphere, allowing the rocket's electronic brain to make thousands of adjustments per second. It's like surfing down a shockwave, guided by incredibly smart metal feathers.

All the while, the engines play a crucial role in this fiery ballet. A 'boostback' burn might kick in high above, cancelling sideways momentum and nudging the booster towards its landing target. Then, hitting the denser air like a wall, a 'reentry' burn acts as a massive airbrake, shielding the base and dramatically slowing the descent before friction can melt it into slag. The grand finale? The 'landing burn'. Just moments before impact, the engine (or engines) relight, throttling with astonishing agility, fighting gravity blow-for-blow, aiming for that perfect zero velocity kiss the instant the landing legs deploy and meet the Earth. It's less like dropping something and more like teaching a needle to land on its point during an earthquake, guided by GPS, gyroscopes, and a river of real-time data.

But landing is only surviving the fall. How do you endure the fire? The inferno of reentry, thousands of degrees hot, is an ordeal few materials can withstand even once, let alone repeatedly. Old rockets wore 'ablative' shields – basically, sacrificial layers designed to char, melt, and carry heat away. Effective, but strictly one-use. The reusable dream demanded something tougher, something resilient.

Enter the unsung heroes: advanced materials, particularly Thermal Protection Systems (TPS). Think spacefaring armor. Materials like ceramic matrix composites (CMCs) are game-changers – lightweight yet incredibly strong, capable of shrugging off unbelievable heat and the shock of rapid temperature swings, launch after launch. Unlike the delicate, high-maintenance tiles of the past (think Space Shuttle), these modern shields, like SpaceX's PICA-X or the tiles cladding Starship, are built for the long haul. They aren't just heat-resistant; they're warriors designed to face the furnace, land, get checked over, and fly again.

And the payoff for all this intricate engineering, this material science wizardry? It's rewriting the economics of space. Imagine if transatlantic flights required a brand new plane for every single trip! That's the barrier reusability is smashing. Where sending a kilogram to orbit once cost as much as a luxury car ($10,000+), reusable rockets are slashing that price dramatically, potentially towards $1,500/kg or even less. Suddenly, the "final frontier" doesn't feel quite so locked behind an impossible price tag. Spreading the colossal cost of the hardware over dozens of flights changes the entire equation.

So, what's the secret recipe, the design philosophy for these cosmic boomerangs?

Teach it to Come Home: Design from scratch knowing the most valuable parts must return safely. Give it the brains (GNC), the steering (grid fins), the throttle control, the landing gear. Recovery isn't an add-on; it's the mission.

Build it Tough, Build it Smart: Use materials that laugh at heat and stress. Design structures that expect to be shaken, baked, and flown again. Make it easy to inspect, easy to trust.

Get Ready to Fly Again, Fast: Think like an airline, not a monument builder. Aim for quick checks, easy component swaps, and smooth operations from touchdown to the next takeoff. Minutes and hours, not months and years.

Weave it In From the Start: Reusability can't be bolted on later. It has to be in the DNA of the rocket, influencing every choice from its shape and fuel to its software and ground support.

This journey – teaching rockets to land on their feet and shrug off the fires of reentry – is more than just clever mechanics. It's unlocking a future where space is not just a destination for the few, but a domain for the many. It paves the way for vast satellite networks connecting everyone, for deeper exploration of our solar system, maybe even for taking a weekend trip to see the Earthrise. These engineering breakthroughs aren't just technical marvels; they are the keys to unlocking humanity's future beyond the cradle.

From Moonshots to Marketplaces: How We're Building Our Future Among the Stars

Remember when space was a silent movie? Black and white images of flags on the moon, the crackle of mission control, a story told mostly by governments sketching lines across the cold void. It felt distant, almost mythical. But hold onto your hats, because someone just flipped the switch to technicolor, cranked up the volume, and threw open the doors. The final frontier isn't just for pioneers with national flags anymore; it's rapidly becoming a bustling marketplace, a dynamic extension of our own world – the space economy is booming.

Forget distant dreams; we're talking about a nearly half-trillion-dollar ($400 billion!) reality right now. That's not government spending; that's the hum of commerce echoing in orbit. How did we get here? Not through slow bureaucratic shifts, but through sheer, audacious grit. Think of companies like SpaceX, Blue Origin, Rocket Lab, and a whole galaxy of startups not just knocking on the door, but engineering a new way to blow it clean off its hinges. They looked at rockets – these titans of engineering – and asked, "Why throw them away?" Their answer – reusable launch vehicles – didn't just cut costs; it fundamentally changed the economics of reaching for the stars. They shrunk satellites from the size of buses to loaves of bread, making space accessible like never before. Suddenly, launching your vision into orbit wasn't just a superpower privilege; it was a viable business plan.

And this cosmic construction site needs people. Lots of them. Sure, the brilliant minds designing rockets and deciphering data from distant worlds are crucial – STEM fields are absolutely thriving. But picture this: the welder on the factory floor meticulously joining segments of a future launch vehicle. The software guru crafting the intricate digital ballet that guides a satellite swarm. The technician in a cleanroom, assembling components with near-surgical precision.

But zoom out further! See the logistics wizard orchestrating a supply chain that stretches from a desert launchpad to an orbiting station? The marketer crafting campaigns to bring satellite internet to remote villages? The sharp legal mind navigating the largely unwritten rules of space law? The financial analyst securing the millions needed to fuel these ambitions? This isn't just about astronauts; it's about builders, communicators, administrators, HR pros, safety officers – a whole vibrant ecosystem mirroring our own, creating real jobs and weaving the future of space directly into the economic fabric of life right here on Earth.

What's pouring fuel on this fire? A torrent of private cash. Venture capitalists, once fixated on apps and biotech, have looked up and seen infinite potential. In 2022 alone, a staggering $7.7 billion flowed into space ventures, according to Space Capital. This isn't the patient, often cumbersome funding of governments. This is fast, demanding, innovation-hungry capital. It empowers startups to dream big, fail fast, and pivot quicker. It's the high-octane boost behind everything from constellations beaming internet worldwide, to orbital platforms watching over our changing climate, to the first whispers of manufacturing goods in space, and even planning the logistics for delivering supplies to a future moon base. The result? Innovation isn't just accelerating; it feels like it's hitting warp speed.

So, how do we ground this celestial ambition? How do nations ensure this isn't just a sideshow, but a core part of future prosperity? It's about smart integration. Think of governments not as the sole actors, but as savvy choreographers. They're crafting sensible regulations – rules of the road for space. They're building the essential "on-ramps" like spaceports. They're becoming smart customers, buying services like cargo runs to the ISS or data streams from commercial satellites, creating stability. Critically, it's about bringing space down to Earth. Imagine farmers using satellite eyes to tend crops with pinpoint precision, shipping companies tracking goods seamlessly across continents, or disaster response teams getting life-saving weather data faster than ever. It means nurturing the next generation of talent, ensuring our schools and universities are preparing people for these incredible careers.

This isn't science fiction anymore. The $400 billion space economy, powered by relentless private energy and investment, is generating real value, diverse jobs, and mind-bending innovation today. The grand challenge, and the immense opportunity, is to intelligently weave this powerful new thread into our national economies, ensuring that humanity's next giant leap outward translates directly into tangible progress for everyone back home. The stars are open for business, and we're all invited to participate in building that future.

The Cosmic Crossroad: Navigating the New Rush Beyond Earth

Look up. The quiet expanse isn't so quiet anymore. Humanity's reaching for the stars again, but this time, it's not just superpowers planting flags. It's a buzzing, jostling crowd: dreamers building rockets in garages, billionaires charting courses to the Moon, companies promising orbital vacations and asteroid fortunes. It's exhilarating, a true frontier opening up. But like any frontier, it's wild, untested, and we're scrambling to write the rules of the road while the traffic is already speeding by. Our old maps – the regulations drawn up when space was mostly empty – are struggling to guide this new, complex cosmic dance.

Think about the launchpads right here on Earth, especially in the US. Imagine the engineers, fueled by coffee and ambition, finally ready to send their creation skyward. But first, they hit the regulatory queue. The FAA, the designated traffic cops for space launches, are facing a deluge. Picture stacks of complex applications from SpaceX's star-hopping ambitions, Blue Origin's tourist trips, Rocket Lab's workhorse vehicles, and countless smaller players dreaming big. It's not just paperwork; it's a fundamental tension wire humming between "Let's innovate!" and "Let's be absolutely sure no one gets hurt." As legal eagles writing in space journals point out, these delays aren't just inconvenient; they can stall dreams, scare off investors, and make it harder for the little guys to even get into the game. The conversation isn't just about more funding for the FAA; it's about finding smarter, more flexible ways to ensure safety without strangling the very innovation that's pushing us outward. Maybe it's less about prescribing how to be safe, and more about defining what safety looks like, letting these bright minds figure out the best way to get there.

Now, zoom out, way out, to the global stage. We have this foundational document, the Outer Space Treaty of 1967. Born from Cold War anxieties, it was remarkably wise for its time, declaring space open to all, forbidding nations from claiming the Moon like conquistadors, and making countries responsible for everything launched from their soil. But 1967 couldn't imagine today. As legal minds exploring Air and Space Law regularly debate: What happens when a company, not a country, starts digging up lunar regolith? Is that 'claiming' it? And who's liable when a piece of debris from a long-dead Soviet satellite hits a brand-new commercial one launched by a multinational consortium? The Treaty offers grand principles, like a constitution, but lacks the specific bylaws for cosmic resource management, orbital traffic jams, and cleaning up our celestial backyard – things critical to today's space reality. This fuzziness creates risks, the kind that could spark real conflict as asteroid mining shifts from sci-fi fantasy to boardroom strategy. Efforts like the Artemis Accords are trying to fill these gaps, particularly on using space resources, but it's like a group agreeing on house rules while others aren't part of the conversation – we need everyone on the same page.

And then there's safety – the absolute bedrock. Every astronaut, every engineer knows space is unforgiving. We often learn the hardest lessons through failure. Remember Astra's launch in 2021? That heart-stopping moment shortly after liftoff when an engine quit, the rocket drifted, and the mission failed. What followed wasn't blame, but painstaking investigation. Under the FAA's watchful eye, teams pored over data, examined twisted metal, and re-evaluated every step. It wasn't just about finding the faulty fuel line; it was about understanding why it failed – was it design, testing, assembly, procedure? This cycle – failure, deep dive, fixing the root cause, getting the regulator's green light – is how we inch towards safer spaceflight. It's how we build resilience. Sharing these hard-won lessons, whenever possible, lifts the safety tide for everyone venturing off-planet.

So, where do we go from here? Imagine a future with conflicting space laws in every nation – a chaotic mess that stifles collaboration, multiplies dangers, and potentially turns space into a lawless scrapyard. That's not the future anyone wants. We need harmony. Think about how we manage global air travel with bodies like ICAO. We need something similar for space:

Reaffirming the Core: Dust off the Outer Space Treaty, maybe add some modern clarity through the UN's space committee (COPUOS).

Writing the Rulebook: Get everyone – industry pioneers, cautious regulators, international experts – together to hammer out shared technical standards for managing space traffic, tackling orbital debris (like designing satellites to burn up safely or rewarding cleanup efforts), and operating spacecraft responsibly.

Sharing the View: Create a global "mission control" for sharing data on where everything is, preventing cosmic fender-benders.

Working Together: National regulators need to talk, share notes, streamline approvals for missions that cross borders, and coordinate oversight.

Staying Agile: Technology is leaping forward. Our rules need built-in mechanisms to adapt, keeping pace with the next giant leap.

Building this framework is a massive undertaking, a diplomatic and technical Everest. It means balancing national pride with the shared need for a safe, sustainable space environment. It requires governments to lead, companies to invest beyond just launch success, and thinkers to keep charting the legal and ethical pathways. This isn't just about managing risk; it's about being good ancestors, ensuring the boundless potential of space is available not just for us, but for all the generations who will look up and dream after us.

The Advent of Space Tourism

From Stardust Dreams to Footprints on the Threshold: Humanity Reaches for Space (Again)

For the longest time, space felt like a story whispered only in hushed, hallowed halls. It belonged to the elite few – heroes strapped into capsules, symbols of national pride launched by the immense machinery of governments [Evidence: Think Neil Armstrong, Yuri Gagarin - names etched in history by NASA and Roscosmos, funded by taxpayers]. For the rest of us, gazing up at the night sky, space was pure imagination, a destination locked behind insurmountable walls of cost and training. The dream of floating among the stars, seeing our world as a fragile blue marble, seemed destined to remain just that – a dream.

But dreams have a funny way of catching fire, especially when fanned by bold vision and deep pockets. The turn of the millennium crackled with a new energy. Visionaries, entrepreneurs who'd already reshaped Earth-bound industries, turned their gaze upwards [Evidence: Names like Branson (Virgin Galactic), Bezos (Blue Origin), Musk (SpaceX), and the team at Axiom Space became synonymous with this new space race]. Their goal wasn't just exploration; it was access. They saw the final frontier not as a government reserve, but as the next great human adventure playground. The key? Making the ride cheaper. They poured fortunes into cracking the code of reusable rockets – machines that could fly, return, and fly again, slashing the astronomical price tag of getting anything, or anyone, off the ground [Evidence: Watch a SpaceX Falcon 9 land upright, or a New Shepard capsule float back under parachutes - these aren't CGI, they're the engines of this new era].

Then came 2021. The dream suddenly felt breathtakingly close.

On a sun-drenched July day, Virgin Galactic's sleek spaceplane, VSS Unity, dropped from its mothership and ignited its rocket. On board, Sir Richard Branson, the company's exuberant founder, wasn't just testing hardware; he was living his own lifelong ambition [Evidence: Virgin Galactic's Unity 22 mission logs tell the technical story]. Hurtling to over 80 kilometers, they touched the edge of space. Branson's words upon return weren't technical jargon; they were pure, unadulterated awe. Seeing Earth, he said, was "a life-changing experience," a stark reminder of its beauty and vulnerability [Evidence: You can almost hear the wonder in his voice in post-flight interviews]. For a few precious minutes, they floated, weightless, experiencing that magical, coveted sensation that had been the sole privilege of astronauts [Evidence: Every astronaut description of microgravity echoes this unique feeling]. Unity 22 wasn't just a test flight; it was a declaration: the door to space tourism was creaking open.

Just nine days later, the cosmic baton passed. Blue Origin's New Shepard rocket, a fully automated capsule booster system, roared to life [Evidence: Blue Origin's NS-16 mission archive confirms the flight]. Inside, founder Jeff Bezos shared the ride with his brother Mark, but the real headlines belonged to their companions. Wally Funk, an aviation legend denied her astronaut wings decades earlier simply for being a woman (one of the "Mercury 13"), finally achieved her dream at 82, becoming the oldest person to reach space. Beside her, 18-year-old Oliver Daemen became the youngest [Evidence: Crew details paint a picture of generations touching the void]. Their capsule soared higher, past the symbolic 100km Kármán line, offering stunning panoramas through giant windows before floating back to the Texas desert [Evidence: Onboard footage captures the wide-eyed wonder]. Funk's triumphant emergence from the capsule wasn't just a personal victory; it was a powerful symbol of perseverance, adding a deeply human narrative thread to the technological leap.

But these brief suborbital hops, exhilarating as they were, were just the appetizers. Axiom Space was already cooking up the main course: living and working in orbit, not just peering at it. In April 2022, they didn't just nudge the door open; they walked right through it, booking a ride on a SpaceX Crew Dragon to the International Space Station [Evidence: NASA and Axiom's Ax-1 press kits detail this orbital venture]. Four private citizens – led by a former NASA pro, Michael López-Alegría – weren't just passengers; they were pioneers setting up temporary camp on the ISS [Evidence: Ax-1 docking logs mark their arrival]. Their mission, Ax-1, wasn't a quick joyride; it was over two weeks packed with science experiments and outreach, sandwiched between the sheer wonder of watching continents drift by minute by minute [Evidence: The mission's science manifest shows this was no holiday]. Their accounts spoke of the sublime beauty, the unexpected camaraderie of station life, but also the demanding reality – space adaptation is tough, and their initial schedules were punishingly intense, a real-world lesson in orbital living [Evidence: Post-mission interviews revealed the steep learning curve, even for trained individuals].

The Price of a Heavenly View

So, the dream is touching reality, but let's talk brass tacks: this cosmic vista currently comes with a celestial price tag. Virgin Galactic, after initial lower prices, set their standard ticket at a cool $450,000 [Evidence: Virgin Galactic's investor reports confirm this pricing]. Hundreds have signed up, proving there's a queue of wealthy adventurers eager for those few minutes of weightlessness [Evidence: Reservation numbers cited in company reports show significant demand]. Blue Origin played coy, auctioning its first paid seat for an eye-watering $28 million (though the winner flew later), with subsequent seats likely in the same ballpark as Virgin's, though specifics are guarded [Evidence: Auction results were public; subsequent pricing relies on industry analysis].

Axiom's orbital stays? That's a different universe of cost. Each Ax-1 seat was estimated around $55 million [Evidence: Widely reported figures in aerospace news, reflecting launch, training, capsule, and ISS 'rent']. This reflects the sheer scale of the undertaking – weeks in orbit, complex life support, cutting-edge rocketry. It's the pinnacle experience, with a price to match. Axiom is already booking flights Ax-2, Ax-3, and beyond, signaling this ultra-luxe orbital tourism is here to stay [Evidence: Axiom's own announcements outline their future ISS mission plans].

Making the Dream Reach Further: Sketching Tomorrow's Space Ticket

Right now, "accessible" space tourism feels like a contradiction in terms. But remember the early days of flight? Only the rich could afford those noisy, rickety biplanes. Technology, competition, and scale eventually brought air travel to the masses [Evidence: The history of aviation shows a clear trend of cost reduction and increased accessibility over time]. Space tourism might be on the same trajectory, albeit a much steeper one.

Getting there means attacking the barriers:

Slash the Launch Costs: This is the big one. Relentless innovation in fully reusable rockets (like SpaceX's ambitious Starship or Blue Origin's upcoming New Glenn) is paramount. More flights, streamlined manufacturing, and fierce competition are the ingredients needed to push prices down [Evidence: The R&D focus of major players is squarely on full reusability; basic economics dictates competition lowers prices].

Offer Different Flavors of Flight: Not everyone needs the $55 million ISS package. What about:

Gentle Ascents: Companies like Space Perspective are designing luxurious balloon capsules reaching the stratosphere (~30km). No weightlessness, but hours of breathtaking views of Earth's curve for a projected $125,000 – still costly, but a different league [Evidence: Space Perspective's public materials detail their 'Spaceship Neptune' concept and pricing].

Earth-Bound Thrills: Imagine hyper-realistic VR journeys combined with the G-forces of a centrifuge or the brief weightlessness of a parabolic flight (the 'vomit comet'). These could offer a powerful taste of space for far less, perhaps inspiring future travelers [Evidence: Astronauts already use these tools for training; commercial adaptation is feasible].

Open the Airlock Wider: Current flyers need to be pretty fit. Future designs could tailor experiences for people with different physical abilities or medical conditions. Designing inclusive cabins and flight profiles is the next ethical and practical frontier [Evidence: Early microgravity accessibility studies on parabolic flights show this is being considered].

Think of Virgin Galactic's own journey: from the X Prize-winning concept (SpaceShipOne, 2004), through heartbreaking tragedy (the 2014 crash), years of testing, to finally flying passengers [Evidence: Virgin Galactic's public history and accident reports document this evolution]. Each flight, each passenger's reaction, each challenge overcome (like Ax-1's intense workload informing future planning) is a lesson learned, paving the way.

The stardust dream hasn't fully settled into everyday reality yet. These first flights are like the crack of dawn – spectacular, promising, but still illuminating only a select few. Yet, the trajectory is set. Driven by relentless innovation and that innate human urge to see what's over the horizon, the journey towards a future where looking down on our beautiful, fragile Earth is an experience more can share, has truly begun. The final frontier is opening, one breathtaking, expensive, but utterly human flight at a time.

From Star Charts to Seatbelts: Weaving the Dream of Everyday Spaceflight

Imagine looking up, not just at stars, but at the possibility of touching them. For decades, that dream belonged to a select few – test pilots and scientists strapped into government machines. But the cosmos is whispering a new invitation, shifting from the language of national prestige to the dialect of human wonder. We stand at the dawn of the citizen astronaut, and with it comes a profound challenge: how do we craft vessels not just for reaching orbit, but for experiencing the universe, safely and soul-stirringly? This isn't just engineering; it's architecting awe.

Designing the Divine View: More Than Just a Window

Why do we yearn for space? Often, it's for that view – the breathtaking, life-altering panorama of our fragile blue marble suspended in infinite black. Blue Origin understood this deep-seated desire wasn't a secondary perk; it was the main event. They didn't just add windows to their New

Shepard capsule; they designed the capsule around the view. Imagine stepping inside and finding, not a cramped porthole, but vast panes of engineered crystal – nearly four feet tall! – framing the cosmos like a masterpiece. Each seat is a front-row ticket to the celestial ballet.

But building the universe's ultimate picture window isn't simple. It's a high-stakes balancing act. How do you pierce a pressurized hull with such massive openings without compromising the sheer strength needed to survive the violence of launch and the fiery kiss of reentry? The answer lies hidden in meticulous designs, in patented layers of advanced materials and reinforced structures – feats of engineering that wrestle physics into submission, all so a tourist can gasp freely at the Earthrise. Beyond the glass, the cabin itself is sculpted for wonder, with ergonomic seats that cradle you and automated systems that pilot the ship, freeing passengers from complex tasks. The goal? To let you simply be there, fully present to the profound spectacle unfolding outside.

The Unspoken Promise: Safety Woven into the Fabric

While the experience fuels the dream, safety is the bedrock upon which it's built. Launching untrained civilians into an inherently hostile environment demands a paradigm shift. We can't rely on years of astronaut training; the spacecraft itself must become the ultimate guardian. Redundancy isn't just a good idea; it's a sacred trust. Think of it as invisible safety nets woven from duplicate systems. If one falters, another seamlessly takes over.

Consider something as vital as the air you breathe. Civilian spacecraft don't just have an oxygen tank; they have layers of life support – multiple tanks, independent lines, readily accessible backup masks – a chorus of systems ensuring that precious breath continues, even if one voice falls silent. Engineers constantly play devil's advocate, rigorously analyzing potential failures (what aerospace folks call FMEA) to ensure these life-giving systems are robust against the unexpected.

And what if the unthinkable happens during ascent? Blue Origin's New Shepard features a "pusher" escape system – a powerful motor poised to thrust the entire capsule and its occupants away from a failing booster, a guardian angel built of propellant and quick-fire sensors. It's a different approach than the tower-like "tractor" systems of old, tailored for this new era. Coupled with multi-stage parachutes designed to deploy reliably even with partial failures, the journey back to Earth is cocooned in layers of engineered vigilance.

Charting the Star Lanes: Rules for a New Frontier

Building these incredible machines is one thing; flying passengers regularly is another. How do we ensure these voyages are acceptably safe? In the U.S., the FAA's Office of Commercial Space Transportation holds the reins, licensing these bold ventures. But the rulebook is still being written. Currently, regulations (like the specific rules in 14 CFR Part 460) focus on verifying critical safety systems work and ensuring passengers truly understand the risks – what's called "informed consent."

This isn't like booking an airline ticket, where aircraft are certified based on millions of flight hours. Commercial spaceflight is younger, evolving rapidly. The vehicles haven't accumulated that exhaustive statistical track record. So, while the FAA mandates rigorous testing, safety analyses, and demonstrated escape capabilities, passengers must acknowledge they are pioneers embarking on a journey with inherent, higher risks than hopping on a 747. Finding the right global standards – balancing stringent safety with the need to encourage this incredible innovation – is one of the great ongoing dialogues of our time.

Real Ships, Real Dreams: Learning from the Pioneers

This isn't just theory; it's happening now:

Blue Origin's New Shepard: It's the quintessential suborbital tour bus. Each flight carrying civilians – dozens so far as of early 2025 – is a real-world test of those giant windows, the automated systems, the pusher escape mechanism. Every journey adds invaluable data, refining the blueprint for accessible space joyrides.

Virgin Galactic's SpaceShipTwo (VSS Unity): Taking a different path with its air-launch system – dropped from a massive carrier plane before igniting its rocket – Virgin Galactic also chases the dream of the citizen astronaut. Despite facing and overcoming significant challenges (including a tragic accident that led to crucial safety improvements), its flights offer unique insights into winged reentry and how to prepare everyday people for the physical demands of spaceflight.

SpaceX's Crew Dragon (Private Missions): Originally built for NASA astronauts heading to the ISS, Crew Dragon has already moonlighted, taking all-civilian crews (like Inspiration4 and Axiom missions) not just to the edge of space, but into orbit for days. These missions are a whole different level – longer, faster, requiring docking with a space station. They demonstrate how incredibly complex, professional-grade vehicles can be adapted for private explorers, testing long-duration life support, advanced training protocols, and the intricate dance of orbital operations, paving the way for future private space stations and orbital tourism.

The Takeaway: Engineering Awe, Responsibly

Crafting spacecraft for you and me is a symphony of disciplines. It blends the audacious pursuit of an experience that can reshape one's perspective with the non-negotiable, unyielding demands of safety. From wrestling with the physics of giant windows to weaving intricate webs of redundant life support, engineers are reimagining spaceflight for a new generation. Guided by evolving regulations and learning from every single flight launched by pioneers like Blue Origin, Virgin Galactic, and SpaceX, the path to the stars, while still steep and challenging, is becoming clearer. The dream is being engineered, one rivet, one window, one safe return at a time.

Beyond the Blue: Dreaming Our Way to the Stars, One Billionaire at a Time

Remember when space was about footprints on the moon, whispered names of distant galaxies, and the collective gasp of humanity watching a rocket pierce the sky? That dream hasn't faded, but it's putting on a tuxedo. The final frontier, once the canvas for national ambition and scientific quest, is being tailored into the most exclusive experience imaginable – the ultimate luxury getaway. Forget superyachts and private islands; the new pinnacle of status, adventure, and perhaps even soul-searching, lies suspended in the black, gazing back at the marble of Earth.

The New Argonauts: Chasing Stardust with Platinum Cards

Who are these new explorers? They're the High-Net-Worth Individuals, the millionaires and billionaires whose terrestrial success stories no longer feel like the final chapter. Reports tell us their numbers are growing, a global tribe with the means ($1 million+ in ready assets) to chase horizons most can only dream of. While not every wealthy soul yearns for zero-G, a potent cocktail of adventurous spirit and unparalleled purchasing power fuels this burgeoning market. Think of it: projected billions pouring into space tourism, driven by those willing to pay anywhere from half a million dollars for a taste of weightlessness to tens of millions for an orbital sojourn.

But it's deeper than just cosmic bragging rights. Psychologists talk about the hunger for unique, transformative moments, for pushing personal boundaries towards self-discovery. Space travel offers this in spades, promising the legendary "Overview Effect" – that profound, almost spiritual shift in perspective reported by astronauts, seeing our world fragile and interconnected against the infinite. It's the search for the ultimate peak experience, a yearning etched in starlight, now accessible to those who can foot the astronomical bill.

Building Castles in the Cosmos: Hotels Amongst the Stars?

Today's voyages are exhilarating dashes: suborbital hops offering a breathtaking glimpse (think Blue Origin, Virgin Galactic), or multi-day orbital missions, often linking up with the International Space Station (thanks to pioneers like Axiom Space hitching rides on SpaceX rockets). But the dreamers are dreaming bigger. Imagine checking into a hotel where the view isn't the ocean, but the swirling clouds of Jupiter, or the curve of the Earth itself.

Concepts like Orbital Assembly's Voyager Station paint this future vividly: vast, spinning structures creating artificial gravity, offering not just bunks, but luxury suites, gourmet restaurants, and recreation lounges. It's the audacious idea of making space not just survivable, but desirable, comfortable, even opulent. Sure, the timelines shift, the engineering challenges are monumental, and the funding needed is galactic in scale. Yet, the detailed plans, the phased approach starting with smaller stations – these aren't just fantasies. They're the serious blueprints for orbital hospitality, signaling a profound belief that humanity's future includes luxury addresses off-planet.

Selling the Void: Marketing Infinity

How do you sell a trip to nowhere and everywhere? You don't market it like a beach holiday. You craft a legend. Space tourism leans heavily on the pillars of ultimate luxury: extreme scarcity, pioneering spirit, and life-altering experiences. The "product" is an entire narrative – the rigorous (yet premium) training, the astronaut wings ceremony, the membership in an elite club of celestial travelers.

Brands like Virgin Galactic curate every step, while Blue Origin emphasizes the soul-stirring view from vast capsule windows. They're not selling seats; they're bottling cosmic awe, offering a perspective shift as the core commodity. It taps into deep desires for uniqueness and achievement, leveraging the sheer limitedness of the opportunity as perhaps the most potent marketing tool of all. It's the velvet rope at the edge of the world.

Blueprints for Reaching Heaven: The Business of Space Travel

Different dreams require different ships:

The Edge-of-Space Experience (Virgin Galactic, Blue Origin): Think of it as the ultimate rollercoaster combined with an epiphany. Brief minutes of weightlessness, unforgettable views, a price tag starting around $450,000.

The model hinges on turning this into a repeatable, safe, premium adventure, building a community around the shared experience.

The Orbital Immersion (Axiom Space, SpaceX-enabled): This is the deeper dive. Days spent orbiting Earth, docking with space stations, requiring tens of millions per ticket. Axiom charters the rides (often SpaceX), manages the complex logistics and training, catering to those seeking a profound engagement with space, sometimes even for private research. SpaceX itself, through missions like Inspiration4, shows how philanthropy and private ambition can also punch tickets to orbit.

The Destination Builders (Orbital Assembly): This is the long game – building the infrastructure, the orbiting hotels and habitats. Their success relies on attracting massive investment now for revenues far down the line, betting on the creation of entirely new markets in the sky.

First Flights, First Whispers of the Future:

This isn't just theory. Virgin Galactic has sent its first paying customers to the edge of space. Blue Origin has flown numerous crews, turning celebrities and executives into astronauts for a day. Axiom Space has successfully run multiple private missions to the ISS, proving the market for high-end orbital journeys and testing the waters for their own future station. And SpaceX acts as the powerful engine enabling many of these private dreams, while potentially eyeing its own direct-to-consumer space travel with Starship.

The Takeaway:

The cosmos is calling, and for the first time, individuals with extraordinary wealth can answer personally. This nascent market, blending audacious technology with timeless human desires for exploration and status, is pushing the boundaries of what "luxury" even means. Yes, the challenges are immense – safety, cost, sustainability, regulation. But the trajectory is clear. The final frontier is cracking open, and while it begins as the playground of the ultra-rich, it marks the first steps towards a future where humanity's horizons extend far beyond the familiar blue sky. The dream of space is becoming bookable, one stardust-sprinkled itinerary at a time.

Our Cosmic Backyard: Dreaming Responsibly Beyond the Blue

We stand on the shore of an infinite ocean, the cosmos, and the urge to set sail is woven into our very being. The roar of rockets signals a new dawn of exploration, whispered promises of Martian colonies, orbiting hotels, and satellites connecting every corner of our world. It's exhilarating! But as we reach for the stars, we must look down at our hands – are they clean? Every launch leaves a mark, not just on the launchpad, but high above, painting invisible scars on our atmosphere and cluttering the very space we yearn to explore. This isn't just about engineering marvels; it's about growing up as a spacefaring species, learning to tidy our cosmic room.

The Price of the Heavens: Earth's Atmospheric Sigh

Picture it: the ground trembles, fire pushes metal skyward – a breathtaking spectacle. Yet, this power leaves an atmospheric echo. Take the reliable Falcon 9. Each flight, burning refined kerosene, is like exhaling roughly 300-400 tons of CO_2 straight up. But it's more complex than just CO_2. Kerosene flames release tiny, dark particles of soot, like microscopic embers flung high into the stratosphere. Up there, far from weather that could wash them away, these particles linger, potentially warming the thin air and nibbling at our protective ozone layer – a subtle, high-altitude consequence we're only beginning to grasp. Then there are the solid boosters used by many rockets, puffing out clouds of aluminum oxide. Imagine fine, shimmering dust hanging miles above us, particles whose long-term effects on atmospheric harmony are still a scientific puzzle. As launches become commonplace, fueled by dreams of global internet and holidays with an orbital view, this collective atmospheric sigh grows louder, demanding we find a cleaner way to leap off-world.

The Kessler Cascade: Our Orbital Junkyard Nightmare

Look up on a clear night. Beyond the stars, there's an invisible, high-speed junkyard swirling around our planet. Decades of ambition have left behind celestial ghosts: dead satellites tumbling silently, spent rocket bodies, shrapnel from accidental collisions, even astronauts' lost tools. The European Space Agency counts tens of thousands of large pieces, millions of smaller ones, and over a hundred million flecks bigger than a grain of sand. Sounds small? At orbital speeds – think multiple times faster than a bullet – even a paint chip becomes a deadly projectile, capable of shattering a working satellite. This escalating mess fuels the fear of the "Kessler Syndrome," a chilling concept where collisions create more debris, leading to more collisions, until entire orbital highways become impassable no-go zones, trapping us beneath a dome of our own making. The very real 2009 smash-up between Iridium 33 and Kosmos-2251, which instantly created thousands of new hazards, wasn't just an accident; it was a warning flare, illuminating the fragility of the orbital infrastructure that powers our modern lives – from the maps on our phones to the forecasts on the news.

Charting a Cleaner Trajectory: Whispers of Greener Fire

But despair isn't the answer; innovation is. Brilliant minds are brewing cleaner ways to touch the void. Methane emerges as a front-runner. When burned with oxygen, it's potentially a much tidier reaction than kerosene, leaving behind less of that stubborn high-altitude soot. Plus, it holds a tantalizing sci-fi promise: future explorers could potentially brew methane fuel on Mars using the Red Planet's air and water ice. Imagine rockets refueling themselves light-years from home! Pioneers like SpaceX with their Starship's Raptor engines and Blue Origin with their BE-4 are already pouring resources into making this methane dream a powerful reality. And let's not forget hydrogen – the ultimate clean fuel, producing only water vapor. It's powerful, proven (think Space Shuttle!), but tricky – keeping super-cold liquid hydrogen contained is a constant engineering challenge.

Flying Responsibly: Writing the Rules for the Tourist Trail

The dawn of space tourism – quick trips to the edge of space or even orbital stays – throws down a gauntlet. Can we build this industry sustainably from the ground up? Here's how we start:

Honest Emissions: Let's demand clear labels. Just like cars have mileage ratings, launches need standardized reporting for CO2, soot, and other pollutants. Eventually, we need limits, nudging companies towards cleaner fuels like methane or hydrogen.

No Littering Allowed: Every tourist vehicle, every stage, needs a clear "end-of-life" plan before it flies. Controlled descents to burn up safely, or moving retired craft to designated "graveyard" orbits – leaving junk behind shouldn't be an option.

Rewarding Clean Tech: Let's offer real incentives – tax breaks, research funds, prime launch slots – for companies genuinely investing in greener engines and sustainable practices. Make responsibility the smart business choice.

Pulling Back the Curtain: Companies need to openly report their environmental footprint – emissions, debris cleanup efforts, resource use. Transparency builds trust and drives improvement.

Global Handshake: Space is global. We need international agreements specifically tackling the environmental side of commercial spaceflight, updating old treaties for this new era. Think global consensus on keeping our shared space clean.

Offsetting (Carefully): While we perfect cleaner tech, robust, verifiable carbon offsetting programs designed specifically for rocket launch impacts could be a temporary bridge. But it must be done right, not just greenwashing.

Our Choice: A Sustainable Leap or a Fouled Nest

Reaching for the stars is perhaps humanity's grandest adventure. But like any explorer entering a pristine wilderness, we have a profound responsibility to tread lightly. The smog of our launches and the glittering garbage belt we're creating aren't distant concerns; they are here, now. By championing cleaner propulsion, driven by bold science, and enacting thoughtful rules, especially as tourism takes off, we can choose a path where exploration doesn't mean exploitation. Protecting Earth and the space around it isn't just good policy; it's ensuring the dream of the final frontier remains alive and accessible for all the generations who will look up in wonder after us. It's simply the right thing to do.

Colonizing New Worlds

Mars: More Than Dust and Dreams – Our Footprint on Forever?

For longer than we've had telescopes sharp enough to truly see it, Mars has been a siren song in our collective consciousness. That rusty-red ember hanging in the velvet black isn't just a planet; it's a question mark, a dare, a persistent whisper of what if? It's morphed from a canvas for wild tales into the ultimate engineering puzzle box, a potential lifeboat for humanity, and maybe, just maybe, the grandest expression of that restless, gotta-see-what's-over-the-next-hill spirit hardwired into us. This isn't just about planting a flag; it's about planting ourselves, and the journey from sci-fi fantasy to achievable reality is being paved, rivet by rivet, by audacious ambition and relentless innovation.

The Starship: Our Interstellar Covered Wagon

Imagine the sheer, neck-craning scale of it. SpaceX's Starship isn't just another rocket; it's conceived as the workhorse, the heavy lifter, the cosmic moving van designed to haul humanity's hopes (and hardware) across the void. Picture this two-stage titan, taller than anything flying today, built not for a single glorious blaze of glory, but to commute. The magic word here is reusability. Like figuring out how to use an airplane more than once, SpaceX bets this is the key to unlocking space travel's padlock, potentially slashing the astronomical costs that have kept us tethered close to home.

But what can it carry? Forget cramped capsules. We're talking about hauling the foundational stuff of a new world – potentially 100 tonnes, maybe even 150, into Earth's orbit first. Then comes the truly clever bit: orbital refueling. Think of it like multiple tanker trucks meeting a long-haul freighter mid-ocean to top up the tanks for the really big journey. Several Starship tankers would rendezvous with the Mars-bound ship, filling its belly with the fuel needed for the months-long trek. This isn't just sending a rover; this is sending the construction site: the habitats, the power grids, the labs, and eventually, the pioneers themselves.

NASA's Cosmic Detective Work: Reading the Martian Diary

While SpaceX builds the ride, NASA and its partners are deep in the vital reconnaissance phase. Their Mars Sample Return mission is less about immediate settlement and more about essential due diligence. Imagine buying a house – you'd want a survey, right? MSR is the ultimate planetary survey. The Perseverance rover has been diligently drilling and caching little vials of Martian rock and soil, geological time capsules. Why bother bringing them all the way back here? Because locked within those grains are secrets about Mars' watery past, whether life ever flickered there, and critically, what lurking hazards might await humans. Is the pervasive red dust merely annoying, or genuinely toxic? Are there unexpected chemical booby traps in the soil?

Sure, the mission's hit some turbulence – budget realities and mind-bending technical hurdles meant a rethink of the how in late 2023 and 2024. Maybe the initial multi-spacecraft relay race was too complex. But the why remains cast in stone: getting pristine pieces of Mars back to Earth labs is seen as non-negotiable homework before we commit boots to the ground long-term. It's the ultimate environmental impact assessment, informing everything from the spacesuits we design to the walls we build to shield us.

Building a Home on the Red Frontier: The Ultimate IKEA Project

Setting up shop permanently on Mars? That's not just a trip; it's arguably the most complex logistical ballet humanity has ever attempted. It's about turning an alien landscape into a place where people can live, work, and maybe even thrive, millions upon millions of kilometers from the nearest hardware store. The roadmap feels like something out of saga:

The Robot Vanguard & Supply Drops: First come the machines. Missions like MSR and other robotic scouts map the terrain, sniff out vital resources like buried water ice (the Martian gold!), and test-drive critical tech. Then, the Starship freighters (or their equivalents) arrive, uncrewed, delivering the "flat-pack" essentials: inflatable habitats, solar panels (maybe later, small nuclear reactors for reliable power), oxygen generators, rovers, tools – the starter kit for Martian living.

Living Off the Land (Because We Have To): Forget supply runs for water and air! True sustainability hinges on In-Situ Resource Utilization (ISRU). We need machines that can suck in the thin, CO2-rich Martian air and crack it open for oxygen (to breathe and make rocket fuel). We need drills that can tap into subsurface ice for water. NASA's little MOXIE experiment aboard Perseverance proved this isn't fantasy – it successfully brewed oxygen from Mars air, a tiny proof-of-concept with enormous implications. Think of it as learning to drink from Martian streams and breathe Martian air (processed, of course!). Future settlers will depend on industrial-scale versions of this.

First Footprints & Raising the Roof: Once the basic infrastructure is robotically deployed and checked out, the first human crews arrive. They aren't tourists; they're construction crews, scientists, and technicians rolled into one. Their first jobs: assembling those pre-delivered habitats, firing up the life support, starting the first Martian gardens (likely glowing hydroponic labs shielded from radiation), and keeping everything running. They'll need serious protection from cosmic rays – perhaps burying habitats under meters of Martian soil (regolith) or using thick, specially designed walls.

From Outpost to Proto-City: More ships bring more people, more diverse skills, bigger tools. The settlement grows organically. Food production scales up. ISRU gets more sophisticated. Maybe rudimentary manufacturing begins, using Martian materials. The goal? A place that mostly sustains itself, a tiny, fragile bubble of Earth life adapting to its new home, though forever linked to Earth by data streams and the occasional arrival of new faces and specialized gear.

The path is staggeringly difficult. We face technological chasms, the brutal toll of radiation and isolation on the human body and mind, and costs that make eyes water. Yet, with tools like Starship taking shape in fiery test flights and missions like MSR preparing to unlock the planet's secrets, the impossible dream feels... less impossible. That roadmap, etched in ambition and fueled by ingenuity, is slowly being drawn. Every successful rocket burn, every gram of data beamed back, every small victory against the harsh Martian environment pulls that distant red horizon just a little bit closer, turning the whisper of what if into the determined murmur of when.

From Footprints to Foundations: Dreaming a Home on the Moon

For generations, the Moon hung in our sky like a silent, silver promise. We touched it, planted flags, gathered rocks, and then... we mostly left it alone. But something's shifting in our cosmic soul. That distant orb isn't just a place to visit anymore; we're starting to imagine staying. We're sketching blueprints for neighbourhoods where dusty footprints become permanent foundations.

Why now? Well, partly because our rockets are getting better, but mostly because hauling everything we need from Earth – every screw, every water bottle, every building block – carries an absolutely eye-watering price tag. Think of trying to build a house by mailing every single brick across the ocean – that's the kind of cosmic expense we face launching things to the Moon. It's forcing us to get clever, to look down at our feet once we land and ask: "What can we make with this?" It's the ultimate pioneering challenge: learning to live off the lunar land.

Moon Dust Dreams: Building with What's Underfoot

Imagine the Moon's surface: not dead rock, but potential. That endless expanse of grey dust and grit – regolith – isn't just debris; it's the Moon's own DIY building kit. Picture it as cosmic concrete mix, just waiting for us to add the ingenuity. This isn't science fiction; it's the heart of concepts like Europe's "Moon Village" – a vision of an open, bustling lunar settlement where nations and companies work side-by-side, perhaps sharing structures literally baked from the ground they stand on.

How? Think cosmic pottery on a grand scale. Scientists are buzzing about focusing sunlight like a giant magnifying glass, or using microwaves, to heat the regolith until the particles fuse together – essentially sun-forging or microwave-baking sturdy bricks and beams without needing glue shipped from Earth. Imagine robotic artisans 3D-printing entire habitats, layer by dusty layer, maybe using a tiny bit of binder or just cleverly melting the regolith itself. Tests show this "Moon-crete" could be strong enough for landing pads that don't kick up killer dust clouds, roads connecting fledgling outposts, cozy habitats, and vital shields against cosmic radiation. The sheer weight we could save by not launching these materials is staggering – potentially freeing up over 90% of the cargo space for other vital supplies, finally making lunar living affordable.

The Helium-3 Question: Earth's Power Plant on the Moon?

But the Moon might hold more than just building supplies. For billions of years, the sun has blasted our neighbour with solar wind, embedding a rare treasure in its dusty skin: Helium-3. Here on Earth, it's incredibly scarce. On the Moon? Potentially millions of tonnes lie waiting. Why the excitement? Helium-3 is the dream fuel for a cleaner kind of nuclear fusion. Imagine power plants providing vast energy for Earth, but without generating mountains of long-lived radioactive waste. It's a tantalizing glimpse of a possible energy revolution, powered by the Moon.

Hold on, though. This dream comes with asterisks the size of craters. Making Helium-3 fusion work demands far more extreme conditions than the fusion reactions we're currently struggling to tame on Earth – it's a massive technological leap. And finding it? It's like panning for gold dust in an ocean of sand; Helium-3 exists in tiny concentrations. Extracting just a little bit would mean processing truly mind-boggling amounts of regolith, heating it intensely (around 700°C!) to coax the precious gas out. It's an industrial challenge on a scale we've never attempted off-world.

First Steps: Water, the Real Lunar Gold

So, while Helium-3 remains a shimmering, long-term goal, the practical path to lunar living starts with something far more fundamental: water. We now know that frozen water – potentially vast amounts – lies hidden in the permanent shadows of craters near the Moon's poles, areas untouched by direct sunlight for eons. Finding and using this ice is the real game-changer right now.

Why? Because water is life. It's air to breathe (split into oxygen and hydrogen). It's rocket fuel on tap (hydrogen and oxygen again). Being able to source water locally slashes our dependence on Earth, making long-term stays truly feasible. This isn't just theory; it's a cornerstone of NASA's Artemis program, aiming to establish a sustainable human presence. Missions are being designed, like the VIPER rover concept, specifically to map out these frozen reservoirs. Commercial pioneers, backed by programs like NASA's CLPS, are racing to build the landers, drills, and processing plants to turn this lunar ice into usable resources. International players, like China with its Chang'e missions, are also actively exploring how to unlock the Moon's bounty.

These first steps – tapping into water ice, practicing construction with Moon dust – are everything. They're the crucial groundwork, the first harvests that prove we can make a go of it up there. They lay the path towards a genuine lunar economy, perhaps one day supporting the grander visions like Helium-3 mining. It's how the Moon transforms in our story, from a place we briefly visited to a vital, humming outpost – humanity's next frontier lodge, and maybe, just maybe, a key to our future.

Subtopic 3: Breathing Life into the Void – Our Cosmic Homestead

To dream of Mars beneath our boots or a city shimmering under the Moon's stark sky is to dream of carrying home with us. Not just the soil or the seeds, but the very essence of what keeps us alive: the air we gasp, the water that quenches, the invisible shield that guards us. Out there, in the silent, stunning emptiness, Earth's generosity ends. No atmosphere welcomes our lungs, no rain fills our cups, and the universe throws radiation like daggers. To survive, let alone thrive, we must become architects of miniature Earths, masters of recycling life's essentials, guardians of the fragile human mind, and ultimately, builders of homes where technology cradles our humanity.

The Endless Cycle: Turning Waste into Wonder

Imagine floating millions of miles from the blue marble, where every exhale, every drop of sweat, every sip taken is a resource too precious to lose. Forget Amazon Prime for water bottles; out here, survival hinges on becoming the ultimate recyclers. This isn't just plumbing; it's alchemy performed by machines like the International Space Station's Environmental Control and Life Support Systems (ECLSS).

The Water Whisperers: Think of the ISS's Water Recovery System (WRS) as a miracle worker. It patiently gathers the ghosts of liquids past – the humidity from breath and sweat, the remnants of hygiene, even urine – and through a technological dance of filters, reactors, and distillation, coaxes them back into pristine water. We're talking 93%, sometimes even 98% recovery [Source: NASA Tech Demo Missions, ISS WPA data]. This isn't just reclaimed water; it's often purer than what flows from taps back home [Source: NASA ISS water quality pubs]. The astronaut sipping coffee today might be drinking the very same water molecules, transformed, tomorrow. Annually, this system gifts back around 6,000 liters, severing a thick strand of the costly umbilical cord to Earth [Figure: NASA ISS Program Science Office]. It's a testament to engineering, yes, but also a profound act of trust in the cycle.

Crafting Air from Water: Oxygen, the very rhythm of our lives, is conjured mainly by the Oxygen Generation Assembly (OGA). It performs a kind of molecular magic, splitting reclaimed water (H_2O) into the hydrogen (H_2) we don't breathe and the vital oxygen (O_2) we do [Source: NASA ISS Systems docs]. But what about that leftover hydrogen? Initially tossed overboard like cosmic breadcrumbs, ingenuity like the Sabatier system now captures it. This clever device marries the waste hydrogen with the carbon dioxide (CO_2) astronauts exhale, forging methane (a potential fuel) and, beautifully, more water [Source: NASA Sabatier research papers]. It's not just waste reduction; it's tightening the knot on the loop, pulling us closer to true self-sufficiency.

The Green Frontier: Inviting Life into the Loop

As elegant as these machines are, they're complex beasts needing constant care and feeding (filters, spare parts). The dream, the next horizon, is to invite life itself into the partnership – bioregenerative life support (BLS). Imagine a habitat humming not just with machines, but with the quiet work of microbes, algae, and plants.

The Living Machine (MELiSSA): Europe's MELiSSA project paints this picture vividly: a chain of life where microbes diligently break down waste, vibrant green algae like Spirulina feast on our exhaled CO_2 while puffing out oxygen and offering themselves as food, and higher plants join the chorus, scrubbing the air, purifying water, and gracing the table with fresh produce [Source: ESA MELiSSA docs/pubs]. Research shows just how eagerly certain algae convert breath into biomass and oxygen under the right conditions [Source: Scientific papers, e.g., Life Sciences in Space Research].

Gardening at the Final Frontier (CELSS & VEGGIE): NASA's explorations (CELSS program) proved we could cultivate Earthly comforts – lettuce, radishes, wheat – under simulated alien suns [Source: NASA CELSS reports]. On the ISS today, the VEGGIE experiment isn't just about supplementing diets; it's about the psychological nourishment of tending a tiny garden, the crunch of a fresh leaf offering a sensory anchor to a distant home [Source: NASA VEGGIE logs/feedback]. These green systems promise more than just resources; they offer resilience, independence, and a touch of the familiar in the profoundly alien.

The Landscape Within: The Human Spirit in Isolation

Building a perfect machine-world isn't enough if the humans inside unravel. Mars is years away, a journey steeped in confinement with a handful of souls, wrestling with communication delays that stretch conversations into echoes, and living constantly under the shadow of risk. The greatest uncharted territory might be the human psyche.

Earth's Deep Freeze University (Antarctica): Where on Earth can we practice for such profound remoteness? The desolate beauty and crushing isolation of Antarctic research stations, especially during the sunless winter-over, offer the closest simulation. Places like Concordia, perched high on the ice plateau, plunge crews into an environment of cold, dark, and confinement eerily mirroring a planetary outpost [Source: ESA Concordia analog research].

Echoes from the Ice: What lessons whisper back from the polar night? Studies consistently reveal the strain [Sources: Acta Astronautica, Aerospace Medicine and Human Performance]:

Sleep's Fragile Rhythm: Inner clocks go haywire without the sun's familiar cues.

Moods Like Shifting Ice: The "winter-over syndrome" speaks of creeping tension, fatigue, and shadows in the mind.

Thoughts Adrift: Monotony and stress can sometimes dull the sharp edges of cognition.

The Friction of Closeness: Small groups, immense pressure – a recipe for sparks to fly.

Fortifying the Mind: This hard-won knowledge shapes how we prepare astronauts: careful selection, rigorous training in navigating stress and conflict, thoughtful scheduling, habitats designed with private nooks and communal hearths, virtual escapes, windows (real or virtual) onto vastness, and vigilant care for mental well-being [Source: NASA Human Research Program]. Analogs like HI-SEAS in Hawaii, simulating Mars life down to the frustrating communication lag, further refined our understanding of keeping teams cohesive and effective when Earth feels like a distant star [Source: HI-SEAS publications].

Building Sanctuaries Among the Stars

All this knowledge converges on a single, monumental task: designing habitats that are more than shelters. They must be self-sustaining ecosystems, cradles for human life on alien shores.

The Integrated Ark: Future Moon or Mars bases won't be collections of parts, but interwoven systems. Advanced life support (likely a blend of machine and nature), robust shielding against cosmic rays (perhaps using local soil or ice), reliable power, and ingenious waste handling must work in concert [Source: NASA Moon to Mars plans, habitat design papers].

Living Off the Alien Land (ISRU): True independence means using what's there. We're already practicing: NASA's MOXIE breathed Martian air and exhaled oxygen [Source: NASA JPL MOXIE results]. Future pioneers will mine water ice locked in craters or soil, 3D-print structures from local dust (regolith), and maybe even extract useful minerals [Source: ISRU research papers]. It's the ultimate homesteading challenge.

Designing for the Soul: The lessons from Antarctica are blueprints for well-being. Habitats need private sanctuaries and welcoming communal spaces. They need light, real or simulated, and views to remind us of the universe outside. Exercise and recreation aren't luxuries; they're essential maintenance for the human spirit. And those green, living modules? They pull double duty, cleaning air and feeding souls [Source: Human factors/architectural research].

Echoes and Futures:

We've learned from ambitious attempts like Biosphere 2, which, despite its struggles, offered priceless, complex lessons in closed systems and human dynamics [Source: Biosphere 2 analysis pubs]. We're building stepping stones like the Lunar Gateway to test deep-space living technologies [Source: NASA Artemis docs]. And remarkably, the struggle to survive off-world sharpens tools for Earth: advanced water purifiers for disaster zones, air scrubbers for confined spaces, and ultra-efficient indoor farms transforming agriculture [Source: NASA Spinoff].

Ultimately, stepping onto another world demands a dual mastery: engineering the perfect life-sustaining bubble, and nurturing the resilient, fragile human spirit within it. From the intricate plumbing recycling yesterday's coffee on the ISS, to the profound quiet of an Antarctic winter, to the grand designs for Martian cities – we are painstakingly gathering the wisdom, the technology, and the courage needed to carry the spark of humanity into the vast, silent cosmos.

Beyond the Blue Marble: Crafting Souls for the Starry Expanse

We're teetering on the edge, aren't we? Ready to leap from the cradle of Earth and plant our fragile footprints on the Moon, Mars, maybe even worlds beyond our wildest imaginings. But building habitats is the easy part, relatively speaking. The real challenge? Weaving the very fabric of society in the stark, echoing silence of space. It's not enough to build shelters; we need to draft blueprints for souls, for communities huddled together under alien skies. How will we live together when home is a distant, beautiful memory?

Old Laws, New Worlds: Can Earth's Rules Reach the Red Dust?

We have treaties, grand documents like the Outer Space Treaty of '67, penned when space was a frontier for flags and footprints, not families and futures. It declares space the "province of all mankind," a noble sentiment ensuring no nation plants a flag and claims Mars as "New [Insert Country Here]." But it leaves pioneers in a bind. How do you build a life, a settlement that lasts, if you can't truly claim the patch of icy ground needed for water, or the mineral vein vital for survival? It's like trying to build a house on land no one is allowed to own or even reliably manage [Source: UN Outer Space Treaty].

Antarctica offers a whisper of a solution. Down there, amidst the ice and blinding white, nations agreed to pause their land claims, focus on science, and keep the peace [Source: Antarctic Treaty Secretariat]. It's worked, mostly. But Antarctica has penguins, not people trying to raise children generations from Earth. Decisions are made slowly, by consensus back home – a luxury a Martian outpost facing a sudden dust storm or equipment failure simply wouldn't have [Source: Analysis comparing ATS/Space Law, e.g., von der Dunk]. You can't wait months for a committee vote when the oxygen levels are dropping.

Today, we see new ideas emerging, sometimes clashing. America's Artemis Accords suggest "safety zones" and rules for digging up lunar resources, feeling pragmatic but perhaps sidestepping that global "mankind" ideal [Source: NASA Artemis Accords]. Europe dreams of a collaborative "Moon Village," a bustling international hub [Source: ESA Moon Village]. It's the age-old tension: do we share this new frontier as one human family, or do we let the familiar forces of national interest and commerce carve it up? Perhaps the answer lies somewhere in between – global rules for the big stuff (peace, protecting alien life if we find it), but letting the Martians, the Lunarians, figure out the day-to-day rules of the road, maybe under a charter they write themselves. But the thorniest question remains: if someone commits a crime millions of miles away, whose laws apply? Who plays sheriff on Mars? [Source: Space Law Jurisdiction Papers].

The Echo Chamber of Isolation: How Space Will Change Us

Forget the legal texts for a moment. Think about the people. Cramped quarters, utter reliance on technology for every breath, the same faces day after unrelenting day, the crushing weight of isolation, and the thin hull holding back deadly vacuum or radiation. These aren't just engineering problems; they are crucibles that will forge entirely new cultures.

We've seen glimpses on Earth: the tight-knit, quirky societies of Antarctic winter crews, the unique jargon and bonds formed in submarines, the resilient cultures of remote Arctic villages [Source: L.A. Palinkas research; isolated group ethnographies]. Tiny groups, under pressure, develop their own ways of being, their own dialects, their own unspoken rules. Intense loyalty can bloom, but so can friction that grates relationships raw [Fact: Antarctic studies track mood/conflict shifts]. Remember Biosphere 2? A simulated Earth colony torn apart not by failing tech, but by human conflict [Source: Biosphere 2 reports]. We must design the social dynamics, not just the life support.

And who gets to go first matters immensely. Those first few pioneers? Their values, their skills, their quirks – that's the cultural DNA they'll pass down. It's the "founder effect" writ large across the solar system [Source: Population genetics/anthropology principles]. Imagine a Mars founded by stoic engineers versus one founded by artists and philosophers. The very soul of the planet could diverge based on that initial seed.

Then there's the view. Seeing Earth, a fragile blue marble hanging in the black, changes astronauts. Frank White called it the "Overview Effect" [Source: Frank White's work]. Could living permanently under that gaze, knowing your species' only home is that vulnerable sphere, fundamentally shift human values? Could it birth societies fiercely dedicated to cooperation, sustainability, maybe even a kind of planetary reverence? Or will the constant struggle for survival breed fear, hierarchy, and tight-fisted control?

Drafting Freedom on Mars: A Constitution for a New World

So, let's imagine drafting a Martian Constitution. It wouldn't just be tweaking Earth's laws; it would be breathing life into a new way of being human, shaped by red dust and distant sun.

Its Heartbeat (Preamble): Forget lofty pronouncements about history. This would likely speak of survival, of shared reliance on the humming machines, of a destiny intertwined with this new, harsh world, and a connection – perhaps loving, perhaps strained – with the Earth left behind.

Life's Essentials (Core Rights): Freedom? Yes, but maybe not the freedom to wander into a lethal dust storm. The "right to breathable air," guaranteed by technology and collective effort, becomes paramount. Privacy? A tricky concept when your every heartbeat might be monitored for safety within the habitat [Analogue: Smart city data debates].

Who's in Charge? (Governance): Early days might demand a "ship captain" model – a skilled leader, maybe elected, focused laser-like on keeping everyone alive [Parallel: Naval/research station models]. But there must be a pathway, as the colony finds its footing, towards something more democratic, perhaps powered by the instant connectivity of a small, wired population.

Sharing the Scarcity (Resource Management): This isn't abstract; it's about life itself. Rules for who gets how much water ice, energy, recycled air, and space within the dome are non-negotiable [Fact: Mars Reconnaissance Orbiter shows water ice is patchy]. Maybe everything vital is held in trust for all Martians.

Sacred Ground (Environmental Protection): We must promise not to contaminate Mars with Earthly microbes if native life exists, and equally, not bring Martian unknowns back to Earth. This needs constitutional weight, echoing our existing planetary protection promises [Source: COSPAR policies].

The Umbilical Cord (Relationship with Earth): Are you Martian, Earthling, or both? How does trade work? Who owns the scientific discoveries? How do we settle arguments between Mars and, say, a corporation back on Earth? Defining this link is crucial.

Built to Change (Amendment Process): They won't know everything on Day 1. The challenges will be alien, unpredictable. The constitution must be adaptable, able to learn and evolve as Martian society itself does.

The Threshold

Figuring out how to govern ourselves beyond Earth is more than just law or engineering. It's a profound test of our humanity, our wisdom, and our creativity. It forces us to draw on everything we know – from Antarctic treaties to the sociology of lonely outposts – and to imagine futures unlike anything we've experienced. If we don't think about this now, proactively, we risk exporting our old conflicts and failings to new worlds, or creating brittle societies that shatter under pressure. Crafting just, resilient ways to live together among the stars isn't just planning; it's about ensuring humanity has a future worth reaching for. It's about writing our destiny not just in the stars, but among them.

Harvesting the Cosmos: Space Mining

Our Cosmic Backyard: More Than Just Sparkly Lights

Look up tonight. That vast, dark ocean dotted with stars? We used to think of it as beautifully empty. But imagine it differently. See those faint specks – the asteroids, our own Moon – not just as distant rocks, but as hidden springs, celestial hardware stores, and treasure chests brimming with materials rare and vital. We're standing on the shore, dreaming of sailing that ocean, and realizing we don't have to pack everything for the voyage. The provisions are already out there, waiting in these "cosmic resource reserves." Learning to find and use them isn't just science; it's the key to truly leaving the nest.

Asteroids: Water Fountains in the Void & Mountains of Metal

Think of certain asteroids, the dark, unassuming C-types, as nature's ultimate canteens. They're not sloshing with water, but they hold it locked deep within their minerals, like cosmic sponges. Some might be nearly a fifth water by weight! Now, why is that ancient, space-borne H_2O so exciting? Because it's the ultimate multi-tool. Zap it with electricity (electrolysis – a kind of space magic!), and you get hydrogen and oxygen. Boom: rocket fuel. Imagine future explorers pulling up to an asteroid "watering hole" to top up their tanks, instead of hauling every last drop from Earth's heavy gravity. It's the difference between a grueling cross-country trek carrying barrels of water, and finding fresh springs along the way. This water isn't just fuel, either. It's air to breathe, shields against harsh radiation, and the lifeblood for growing food far from home.

But asteroids aren't just about quenching thirst. Others, like the metallic giant 16 Psyche that NASA's mission is zooming towards, are like glittering mountains of metal. Psyche is special – scientists think it might be the naked, exposed core of a baby planet that never fully formed. Studying it is like getting a peek into the heart of worlds. But beyond the pure science, it represents a different kind of cosmic fortune. These metal worlds are packed with iron and nickel, sure, but they might also glitter with platinum, palladium, rhodium – metals incredibly scarce on Earth's surface because they sank to our core long ago. While actually mining an asteroid for these treasures is still the stuff of sci-fi dreams (fraught with mind-boggling challenges and wildly speculative price tags – think quintillions!), the knowledge Psyche sends back is priceless. Is it solid metal? A gravel pile? Are the valuable bits clustered or scattered? Knowing this turns vague dreams of space resources into real engineering problems we can start to solve. Psyche is drawing the first detailed treasure map.

Our Moon: An Icy Reservoir Next Door

Closer to home, our familiar Moon held a secret for ages. We thought it was bone-dry, a dead, dusty world. Then, missions like NASA's Lunar Reconnaissance Orbiter (LRO) rewrote the story. Its clever instruments – using lasers, sensing faint starlight, and bouncing radar signals – peered into the eternally dark, frigid craters near the lunar poles. Places sunlight hasn't touched in billions of years. And there, mixed with the lunar soil (regolith), they found it: water ice. Not giant skating rinks, but perhaps billions of tons of frozen water trapped in the cold shadows. Like discovering a massive underground reservoir in our own backyard! This lunar ice offers the same incredible potential as asteroid water: drinking water, breathable air, and, critically, locally-made rocket fuel for the next leap outwards, perhaps to Mars.

Charting the Heavens: Making Maps for the Space Age

Knowing these treasures exist is one thing. Knowing exactly where the richest deposits are, how deep they lie, and how easy they are to get? That's the real challenge. This is where the explorers become mapmakers. Data pouring back from LRO, from asteroid samplers like OSIRIS-REx (which found water-hiding minerals on Bennu!) and Hayabusa2, and soon from Psyche – it's all being woven together. We're creating cosmic geological surveys, treasure maps for the final frontier.

Why This Matters Right Now: Dreams Taking Shape

This isn't just future fantasy; it's shaping plans today:

Artemis Campgrounds: Why is NASA aiming for the Moon's South Pole for its Artemis base camp? Because the maps point there! Proximity to those shadowy craters potentially holding accessible water ice is a huge draw. Upcoming rovers like VIPER are the scouts, sent ahead to drill and "taste" the icy soil, confirming the map's promise. They're checking if the "well" is really where we think it is.

Asteroid Rendezvous 2.0: What Psyche tells us about its metallic heart will dictate how we approach similar asteroids in the future. It informs the design of probes, drills, or robotic arms needed for closer study or, one day, resource extraction. It helps us shift from "Wow, metal!" to "Okay, how do we interact with this specific kind of metal body?"

The Interplanetary Highway: Forget launching everything from Earth. Imagine a future where spacecraft hop between Earth, a lunar refueling station stocked with Moon-ice-derived propellant, and maybe a strategically placed asteroid "gas station," before heading to Mars. Resource maps are the foundation for planning these future supply lines, making deep space travel vastly more sustainable and affordable. It's about building stepping stones across the void.

Cataloging these cosmic resources fundamentally changes our relationship with space. We shift from being short-term visitors, hauling everything on our backs, to becoming long-term inhabitants, learning to live off the celestial land. The data streaming back from these missions isn't just numbers and graphs; it's the blueprint for humanity's expansion, the guidebook for settling the solar system, powered by the very fabric of the cosmos itself. The universe, it turns out, has left the pantry open for us.

Beyond the Backyard Digger: Crafting the Tools for a Cosmic Gold Rush

Think about mining. Maybe you picture dusty prospectors, heavy machinery rumbling under an open sky, the reliable pull of Earth keeping everything grounded. Now, fling that picture light-years away. Swap the blue sky for the black velvet of space, Earth's gravity for a whisper of pull, and your sturdy boots for... well, nothing. Suddenly, digging for cosmic treasure demands more than just a tougher shovel; it demands robots born from science fiction, grappling with physics that wants to send them spinning into the void.

Meet the Trailblazers: The Weaver and the Ice Pick

First up, imagine a construction foreman that fits in a rocket fairing but builds structures bigger than a bus once it gets to work. That's the spirit of Archinaut. Forget shipping pre-built behemoths; Archinaut is like a cosmic spider, equipped with a 3D printer for a spinneret and robotic arms for legs. It arrives compact, carrying the raw thread – polymers, metals – and then weaves structures right there in the vacuum. Think massive solar arrays unfurling like mechanical blossoms, or the backbone of a future space station sculpted against the backdrop of stars. It's not just building; it's becoming. NASA's nod with funding isn't just a pat on the back; it's a bet that this weaver can spin the infrastructure of our future off-world, maybe even using materials found right there on the Moon or an asteroid – the ultimate cosmic recycling program.

But what if the treasure isn't floating freely, but locked deep beneath an alien surface? You need a different kind of hero. Enter TRIDENT, NASA JPL's tenacious little drill. Don't let its size fool you (it's barely heavier than a packed suitcase!). This isn't your garage drill; it's a multi-talented ice pick designed for the Moon's most forbidding landscapes. It hammers, it twists, it bites through dust as fine as flour and ice harder than granite, all while sipping power. Its mission? To plunge into the permanently shadowed craters near the Moon's South Pole aboard the VIPER rover. Think of it as the point man, the one sent ahead to confirm the whispers and orbital scans suggesting frozen water – a resource that could fuel rockets and quench the thirst of future explorers – hiding just beneath the surface. It's venturing into places colder than anything on Earth (~ -223°C!) to answer the billion-dollar question: is the water really there, and can we get it?

The Zero-G Dance: Why Mining in Space is a Slippery Challenge

On Earth, gravity is the ultimate stage manager. It holds drills steady, makes debris fall neatly away, and keeps tons of rock behaving predictably. Take away that familiar embrace, and chaos threatens.

The Pushback Problem: Imagine trying to drill into a wall while standing on roller skates. Push the drill, and you move, not the wall. That's space mining in a nutshell. Without serious anchors – think robotic grappling hooks, corkscrewing legs digging into the dust – the simple act of drilling could send your expensive robot tumbling end over end. Engineers are sketching franticly, testing ways to get a grip where there's almost nothing to grip.

The Dust Cloud Nightmare: On Earth, dust settles. In microgravity, it floats. Drilling kicks up clouds of microscopic, electrostatically charged grit (especially that clingy, abrasive Moon dust the Apollo astronauts complained about!). This isn't just messy; it's dangerous. It can blind sensors, gum up delicate mechanisms, and coat solar panels, choking the robot's power supply. And if you're drilling for ice? The friction might just turn your precious frozen water into escaping gas before you can even scoop it up – like trying to catch smoke with tweezers. Getting the goodies out requires rethinking everything, maybe using puffs of gas or clever conveyors that don't rely on "down."

Catch Me If You Can: Extracting the actual prize – be it water ice or rare metals – is its own puzzle. How do you efficiently bake ice out of lunar soil without losing it all to the vacuum? How do you sort valuable minerals from worthless rock without gravity's help? It's forcing engineers to dream up entirely new ways to process materials, relying on heat, magnetism, or clever chemistry adapted for the strangest factory floor imaginable.

From Blueprints to Bootstraps: Making Cosmic Resources Real

This isn't just whiteboard dreaming. The tools are being built, the missions are flying, and the stakes are sky-high.

The Lunar Gas Station: VIPER, armed with its TRIDENT drill, isn't just a science mission; it's scouting the location for a potential off-world "gas station." Find the water ice, prove you can get to it, and suddenly you can imagine robots extracting it, splitting it into hydrogen and oxygen, and voila – rocket fuel, made on the Moon! This could slash the cost of exploring the solar system, making Mars trips and asteroid ventures vastly more achievable. It's about turning the Moon from a destination into a launchpad.

Building with Moon Dust: Forget shipping bricks. Pioneers like the ESA are figuring out how robots, perhaps cousins of Archinaut, can use the local "dirt" – the regolith – as building material. Imagine robots melting or sintering moon dust with lasers, or mixing it with binders to 3D-print landing pads, shelters, even radiation shields. It's the ultimate bootstrapping, living off the land on an alien world.

The Asteroid Frontier: Further out, but buzzing with energy, are the asteroid miners. Companies like AstroForge and TransAstra aren't waiting. They're designing robots to chase down asteroids, analyze their worth, and figure out how to extract platinum or water using wild techniques like vaporizing rock with concentrated sunlight. It's high-risk, high-reward, pushing robotics and resource extraction into truly uncharted territory.

So, yes, forging the frontier requires new tools. But it's more than just hardware. It's about robotic ingenuity stepping in where humans can't easily go. It's about wrestling with the weird physics of low gravity and the harshness of alien environments. From Archinaut's elegant constructions to TRIDENT's gritty determination, we're building the machines that will not just explore, but utilize the solar system. It's a monumental task, demanding creativity and resilience, but the prize is unlocking the vast wealth of the cosmos, paving the way for humanity's future beyond Earth.

Imagine looking up at the Moon, not just as a poet's muse or a lover's backdrop, but as a potential gas station, a quarry, a treasure chest. That's the shift the U.S. tried to kickstart with its 2015 Space Act. Think of it like this: back in the Wild West, you couldn't own the whole prairie, but if you staked a claim and dug up the gold, that gold was yours. The Space Act aimed for something similar – telling American dreamers and daredevil companies, "Go ahead, build your space shovels and drills. If you pull valuable stuff—water ice from a lunar crater, rare metals from an asteroid—you get to keep it, sell it, build with it."

The idea? To light a fire under private investors. Let's face it, strapping rockets to mining equipment and flinging it millions of miles isn't cheap. You need the promise of a payoff to justify the colossal risk. And it worked, for a while. Companies with names straight out of science fiction, like Planetary Resources and Deep Space Industries, popped up, dreaming of asteroid fortunes.

But Didn't We Promise Not To Plant Flags?

Here's where the cosmic plot thickens. Decades earlier, back in '67, the world came together and signed the Outer Space Treaty – basically, the foundational rulebook for space. Its core message? Space is for everyone. No nation can stick a flag on the Moon or Mars and call it theirs, not by landing on it, using it, or any other cosmic loophole. It's meant to be the "province of all mankind."

So, critics immediately cried foul. Doesn't letting U.S. companies mine and own space resources, under U.S. law, eventually look a lot like the U.S. claiming chunks of space, just indirectly? The 2015 Act cleverly sidesteps this by saying you own the stuff you extract, not the celestial real estate itself. But can you really suck all the water out of a crater without effectively controlling that crater? Imagine trying to mine an asteroid belt – how do you operate without carving out your own zone, warning others off? Legal eagles have been wrestling with this: where does "using" space (allowed) bleed into "appropriating" space (forbidden)? It's a blurry line drawn in moondust.

And what happens when a Chinese rover and an American mining bot arrive at the same promising patch of lunar ice? Or two private companies, backed by different nations with different rules, target the same platinum-rich asteroid? Right now, there's no interstellar referee, no agreed-upon system to handle claim disputes or prevent a celestial traffic jam – or worse. The 2015 Act opened the door, but didn't build the framework for peaceful coexistence out there.

Echoes of Earth: Will We Scar the Heavens Too?

The dream is dazzling: asteroids packed with metals rarer than gold, lunar soil rich in Helium-3 for clean energy, water ice to fuel future explorers. But pause and listen. Do you hear the echoes of Earth?

We know the story here. Mining has built empires and powered progress, but it's also left deep wounds: mountains leveled, forests erased, rivers poisoned with chemical runoff (remember the Animas River turning toxic orange?). Communities have been uprooted, landscapes irrevocably changed.

Will we be better cosmic neighbors? Could asteroid mining create lethal clouds of high-speed debris, turning orbital pathways into minefields? Could lunar operations kick up so much dust that it blinds telescopes or smothers unique geological sites – places holding secrets about how our solar system, maybe even life, began? We talk about "Planetary Protection," mostly to avoid contaminating other worlds with Earth microbes, but maybe we need "Cosmic Conservation" too. Is it okay to chew up an asteroid or scar the Moon just because nothing lives there? We wrestle with similar questions about drilling in pristine wilderness or dredging the deep ocean floor here on Earth.

The ethical tightrope is this: How do we tap into space's bounty without repeating our worst terrestrial mistakes? How do we ensure the benefits aren't just hoarded by the rich nations or corporations that get there first, but genuinely serve that "province of all mankind"?

Charting a Course Together: An Interstellar Rulebook

Relying on a 60-year-old treaty, written before anyone seriously considered asteroid mining, feels like navigating a spaceship with a sextant. Unilateral laws like the 2015 Act just create uncertainty and potential conflict.

What we arguably need is a new chapter: an Interstellar Resource Treaty. Not to replace the Outer Space Treaty's grand vision, but to fill in the crucial details for this new era. Imagine a global agreement that:

Draws Clear Lines: Finally defines the difference between exploring, temporary use, setting up shop, and actually claiming ownership of resources in space. No more blurry definitions.

Creates Cosmic Traffic Control: Sets up an international body to register claims, issue licenses, and prevent companies or countries from tripping over each other. Think air traffic control, but for asteroid mining.

Establishes "Keep Out" Zones & Green Rules: Defines environmental standards for space operations – how to minimize debris, limit surface damage, and maybe even designate "Space Heritage Sites" (like the Apollo landing spots or scientifically unique areas) as off-limits, like cosmic National Parks.

Shares the Wealth: Figures out a fair way to share the economic spoils, ensuring that nations unable to fund their own space missions still benefit from resources considered part of our common heritage. This is tricky, but vital for global buy-in.

Provides Referees: Builds mechanisms to settle disputes peacefully and fairly when disagreements inevitably arise.

We're seeing baby steps towards coordination with things like the U.S.-led Artemis Accords, where nations agree on principles for lunar exploration. But these are more like agreements between partners, not a universal rulebook, and even they raise eyebrows about "safety zones" potentially becoming keep-out zones.

The technology is leaping forward. The ambition is palpable. But without a robust, globally agreed-upon framework, we risk turning the final frontier into a new arena for conflict and exploitation. Crafting a true Interstellar Resource Treaty will be a monumental diplomatic challenge, balancing national pride, corporate profits, scientific curiosity, and our shared ethical duty. But it seems the only way to ensure that our journey into space elevates humanity, rather than just exporting our terrestrial squabbles and scars.

The Cosmic Fork in the Road: Haul It Back or Build It There?

Imagine humanity standing at the edge of an infinite ocean – space. Below the surface, shimmering in the void, lie untold treasures: water locked in lunar ice, precious metals glinting within asteroids, raw materials scattered across the cosmos like a celestial hardware store waiting to be opened. But reaching them raises a fundamental question, one that echoes the choices of pioneers throughout history: do we wrestle these riches back to the familiar shores of Earth, or do we learn to live off this new land, building our future out there?

The thought of bringing space-borne wealth home is intoxicating. Think platinum, iridium – metals so rare and vital on Earth they feel almost mythical. Prospectors dream of lassoing a metallic asteroid, a flying Fort Knox, and dragging its bounty back. But Earth guards its domain jealously. The sheer energy, the astronomical cost of piercing the atmosphere and landing tonnes of material safely, is like trying to swim upstream against a planetary waterfall. It's a high-stakes gamble, requiring staggering upfront investment to chase treasures that, if returned en masse, could potentially crash the very markets they aim to supply. Right now, hauling bulk materials home feels less like sound business and more like a Herculean feat reserved for the truly priceless, or perhaps, the truly patient. The cost of simply leaving Earth is measured in thousands of dollars per kilogram; coming back laden with goods? That's a whole different level of expensive.

But turn your gaze outwards, and a different path unfolds – one whispered about by visionaries and increasingly validated by pragmatic analysis. What if the true value isn't bringing the mountain to Mohammed, but building Mohammed's house on the mountain? This is the philosophy of "living off the cosmic land" – In-Situ Resource Utilization (ISRU).

Take water ice, hiding in the shadowed craters of the Moon or laced within asteroids. It's not just H2O; it's the key to unlocking the solar system. Crack it open into hydrogen and oxygen, and you have potent rocket fuel, brewed right where you need it. Imagine "gas stations" orbiting Earth or parked on the Moon, filled not with fuel painstakingly shipped from our planet's deep gravity well, but with propellant sourced locally. Suddenly, trips to Mars, the asteroid belt, and beyond become vastly more affordable, sustainable. We cut the colossal chain tethering our ambitions to Earth-launched fuel tanks. McKinsey's crystal ball gazes upon these cosmic watering holes as vital infrastructure for a thriving space economy.

And it's not just fuel. That lunar dust, the regolith blanketing the Moon and asteroids? It's not just dirt; it's potential building material. Instead of launching heavy habitats or radiation shields, why not 3D print them using the ground beneath our astronauts' feet? Bake it, sinter it, mix it – transform local grit into landing pads, shelters, and roads.

Weaving Dreams from Stardust: The Asteroid Blacksmith

Here's where the future gets truly tangible. Picture robotic probes landing softly on a metallic asteroid, not just chipping off samples, but becoming cosmic artisans. They mine iron, nickel, cobalt; process them into powders or wires; and feed them into sophisticated 3D printers. Right there, millions of miles from Earth, they begin to create: essential tools, replacement parts for a ship far from home, even the structural beams of a new orbital outpost.

This isn't science fiction evaporating on contact with reality. The European Space Agency (ESA) is already kneading this dough, experimenting with meteorite fragments (our asteroid stand-ins) to perfect the art of space-based metalwork. They're simulating the vacuum, the harsh temperatures, figuring out how to weld and shape metals layer by intricate layer, far from any terrestrial workshop. They're even exploring how to turn that "worthless" regolith into solid structures, perhaps using concentrated sunlight like a celestial forge. The payoff? Freedom. Freedom from Earth's supply chain tyranny, the ability to repair and adapt on the fly, and the power to construct colossal structures – solar power stations beaming energy to Earth, sprawling research labs – too vast to ever fit in a rocket's nose cone. Challenges exist, absolutely – ensuring purity, mastering microgravity manufacturing – but these are the exciting engineering puzzles of forging a new industrial revolution, not insurmountable barriers.

From Blueprint to Reality: Building Our Future Off-World

The leap now is one of scale. From printing a wrench to printing a habitat. From a lab experiment to a functioning lunar construction yard or an orbital factory. This requires orchestration. Armies of smart, autonomous robots acting as our prospectors, miners, processors, and builders, working tirelessly under alien skies. It demands immense power – vast solar farms unfurling in space, perhaps even compact nuclear reactors providing the lifeblood for this off-world industry.

And the foundations are being laid:

Moon Bricks and Lunar Havens: NASA's Artemis program isn't just about footprints; it's about settlements. Plans explicitly involve using lunar regolith, perhaps 3D printed or baked into bricks, to build those first shelters, landing zones, and protective berms – making a home on the Moon with the Moon itself.

Gas Stations in the Sky: Companies like Orbit Fab aren't just dreaming of orbital fuel depots; they're building the prototypes and signing the contracts. These depots, potentially filled with space-sourced propellant, could become the vital pit stops that extend satellite lifespans and make deep-space journeys routine.

Weaving Giants in Orbit: How do you build a solar power satellite miles across? Or an antenna array the size of a football field? You don't launch it; you build it, piece by piece, in orbit. Using metals refined from asteroids and 3D-printed structural components could be the only practical way to assemble the mega-infrastructure needed for humanity's next giant leaps.

So, while the lure of bringing cosmic riches back to Earth remains a distant, glittering prize, the immediate, pragmatic, and ultimately more transformative path seems clear. The smart money, the engineering breakthroughs like ESA's additive manufacturing work, and the sheer necessity of sustainability are all pointing in the same direction: use it there. By learning to harness the resources of space in space, we're not just finding new materials; we're laying the foundation for a permanent, thriving human presence beyond Earth, rewriting our relationship with the cosmos one 3D-printed part, one tank of lunar-brewed fuel, at a time. The crossroads is before us, and the sign points outwards.

Building Homes in Orbit

Beyond the Blueprints: Crafting Our Cosmic Future, One Room at a Time

We stand on the edge of everything, looking up. Humanity, restless and reaching, dreams of setting foot on other worlds, not just as visitors, but as neighbours. This giant leap isn't just about the fiery roar of rockets; it's about weaving the familiar thread of 'home' into the alien tapestry of space. Enter the space architect, a new kind of pioneer, blending hard-nosed engineering with the resourcefulness of a castaway, sketching out dwellings for the final frontier. Their canvas? The vacuum of space and the dusty plains of Mars. Their tools? Clever design and the very ground beneath alien skies.

Cosmic Origami: Unpacking Our Future

Imagine trying to ship a whole house overseas in a single suitcase. Impossible, right? Sending anything to space faces a similar, albeit more extreme, challenge. Every gram, every cubic centimetre, carries an eye-watering price tag dictated by the brute force needed to escape Earth's pull. The ingenious answer? Think cosmic Lego, or perhaps even space-age origami. We design our future homes in bite-sized pieces – modules – launched snugly within rockets, ready to be clicked together or unfolded dramatically upon arrival.

The most dazzling expression of this "pack small, live large" idea came from the visionaries at the now-departed Bigelow Aerospace. Picture this: a module launched like a tightly furled sleeping bag, which, once in space, inflates like a high-tech balloon into a spacious habitat. Their B330 concept promised a staggering 330 cubic meters –

imagine a cozy studio apartment expanding into nearly a third of the entire International Space Station's volume, all from a single launch! It sounds like magic, but it's rooted in clever layers: an airtight inner skin, a super-strong corset (think materials like Vectran) to hold its shape against internal pressure, and tough outer blankets shielding against the thermal extremes and the relentless pings of micrometeoroids. Though Bigelow's journey ended, the dream inflates onward. Sierra Space is nurturing this legacy with their LIFE habitat, designed to bloom in orbit, potentially linked to their sleek Dream Chaser spaceplane or forming part of new commercial space hubs. And clinging bravely to the ISS since 2016 is the BEAM module, a smaller, real-world testbed, quietly proving that these expandable dreams can indeed hold air, and our hopes, among the stars.

Building Castles from Alien Sand: The Homesteader's Revolution

Shipping everything – every wall panel, every screw, every bag of concrete mix – from Earth to build a lasting town on the Moon or Mars? That's not just expensive, it's practically unsustainable. This forces us to embrace the ultimate homesteader's ethos: In-Situ Resource Utilization (ISRU), or simply, living off the land. The most exciting chapter in this story involves turning alien dirt, the regolith blanketing Mars and the Moon, into building blocks.

Forget the water-hungry concrete we know. Imagine future Martians mixing the local red dust with something unexpected, like molten sulfur, thought to be relatively common there. Recent experiments (heating simulated Martian soil with sulfur until it melts, then pouring or even 3D-printing bricks and structures) suggest this "Martian concrete" could be surprisingly strong, potentially even offering some shielding from cosmic rays. Picture robotic construction crews, tireless 3D printers guided from Earth, scooping up Martian soil, mixing it with locally sourced binder – perhaps sulfur, perhaps something synthesized, perhaps even special glues made by engineered microbes – and patiently printing habitats, layer by dusty layer. Of course, it's not easy. Will the binders hold up in the brutal Martian temperature swings? How do we stop the pervasive, fine dust from gumming up the works? But mastering the art of building with alien dust is the key to unlocking truly affordable, large-scale settlements far from home.

From Orbital Outposts to Planetary Homes

This idea of building piece-by-piece isn't science fiction; it's how we've already built our grandest structure off-world. The International Space Station (ISS) is a sprawling testament to modularity – a decade-long orbital construction project, painstakingly assembled module by module, truss by truss, through countless launches and daring spacewalks. It's a floating city built from pre-fabricated parts designed to connect seamlessly in the void.

And the next wave of space stations – Axiom's luxurious orbital suites, Blue Origin and Sierra Space's ambitious Orbital Reef, Nanoracks and Lockheed Martin's Starlab – are all following this modular playbook. They envision launching core elements first, then adding more specialized pieces over time, allowing them to grow and adapt like living organisms responding to the needs of science, manufacturing, and even tourism.

While you can't exactly mine resources in the vacuum of orbit (yet!), the hard-won experience of assembling complex structures like the ISS, and managing life within them for decades, is invaluable. These orbital lessons in modular living are the essential foundation upon which we'll build surface bases. On the Moon and Mars, under programs like NASA's Artemis, the elegance of modular design will finally merge with the rugged resourcefulness of ISRU. We won't just assemble pre-packed kits; we'll combine them with structures literally raised from the alien ground.

So, these architectural dreams – the expandable habitats packing down small, the schemes for baking bricks from Martian dust – are far more than technical drawings. They are the tangible steps, the clever tricks, the sheer ingenuity required to transform humanity from Earth dwellers into citizens of the solar system. They are the blueprints for belonging, wherever we choose to make our next home among the stars.

Breathing Life into the Void: Our Cosmic Homestead

Imagine drifting, weightless, the blue marble of Earth a breathtaking jewel against the endless velvet black. Out here, miles from the nearest soil, the scent of home isn't just a memory – it's the sharp, green perfume of fresh lettuce growing just feet away. This isn't science fiction; it's the dawn of cosmic homesteading.

Keeping pioneers alive and thriving, whether orbiting high above or setting foot on the rusty plains of Mars, isn't just about rockets and radiation shields. It's about the most fundamental human need: food. Shipping celestial groceries is wildly impractical. The real dream, the sustainable future, lies in coaxing life from the void itself, crafting gardens between the stars.

Floating Gardens: A Taste of Earth

Forget soil. Think of it as nutrient-rich lifeblood, delivered directly to thirsty roots. Hydroponics, that clever art of soil-less farming, becomes our lifeline. On the International Space Station, little patches of Earth bloom under specialized lights. Systems like VEGGIE and its bigger sibling, the Advanced Plant Habitat, are our proving grounds. Astronauts, our spacefaring farmers, have carefully tended crops of tangy mizuna, crisp red romaine, and even peppery radishes.

More than just vitamins, these floating gardens are oases for the soul. Imagine, after months of sterile equipment and freeze-dried meals, snipping a fresh leaf, its texture real, its taste vibrant. Astronaut Scott Kelly spoke of this simple act, the presence of growing green things, as a powerful balm against the profound isolation of space. It's a living link to the world left behind, a splash of familiar color in an alien landscape. Of course, challenges sprout alongside the greens – how do you 'water' without gravity pulling drops downwards? How do plants fare under cosmic rays generation after generation? The answers, meticulously gathered and shared in scientific whispers between labs on Earth and orbit, are writing the instruction manual for future space farmers.

Tiny Green Engines: Breathing Easy with Algae

But our cosmic gardening doesn't stop with leafy greens. Enter the unsung heroes: algae. Picture glowing green tanks, not unlike futuristic lava lamps, humming quietly in a corner. Inside, trillions of microscopic powerhouses like Chlorella vulgaris are working miracles. These tiny dynamos are photosynthesis wizards, potentially scrubbing the carbon dioxide we exhale and pumping out fresh oxygen even more efficiently, volume for volume, than their leafy cousins.

And the magic doesn't stop there. Harvest that vibrant biomass, and you have a superfood concentrate – greed with protein (sometimes over half its weight!), vitamins, and essential fats. Think of it as a nutrient-dense smoothie ingredient, cultivated right next door. Experiments, like the German Aerospace Center's bioreactor tested on the ISS, are figuring out how to keep these microscopic farms running smoothly, paving the way for life support systems where breathing fresh air and finding a protein boost come from the same elegant, living system.

Space Age Alchemy: Turning Trash into Treasure

Living light years from the nearest landfill means 'waste' is a word we must unlearn. Every scrap, every leftover, every breath exhaled is a potential resource. We're already masters of water recycling on the ISS, reclaiming nearly every precious drop from breath, sweat, and yes, even urine. But what about the solid stuff – the empty food pouches, the discarded filters, the inevitable human waste?

This calls for something more radical: plasma pyrolysis. Forget smoky incinerators. Imagine harnessing controlled lightning, an intense, contained plasma torch reaching thousands of degrees. In this low-oxygen inferno, waste doesn't burn; it's molecularly disassembled. Complex trash breaks down into simple, useful gases (syngas – think hydrogen and carbon monoxide, potential fuel or ingredients for making more water) and a sterile, glass-like slag. This isn't just waste reduction; it's transformation. Suddenly, yesterday's garbage could become tomorrow's rocket fuel component, building material, or even radiation shielding. It's the ultimate act of cosmic recycling, driven by necessity and ingenuity.

The Symphony of Survival: Weaving it All Together

These aren't just clever gadgets operating in isolation. Th are notes in a complex symphony – the Bio-regenerative Life Support System (BLSS). It's about creating a miniature, human-tended ecosystem, a bubble of Earth life, humming with interconnected efficiency, far from home.

Picture it: Your breath feeds the algae and the hydroponic lettuce. Their oxygen fills your lungs. Waste water, purified, cycles back to nourish the plants or quench your thirst. Solid waste, transformed by plasma fire, yields energy or building blocks. Nutrients locked in waste find their way back to the floating gardens. It's a closed loop, a dance of biology and engineering mimicking the cycles of our home planet, but miniaturized and optimized for the stars.

Projects like Europe's MELiSSA program, with its ground-based loops of bacteria, algae, and plants, are like rehearsals for this grand performance. Even the ambitious, challenging Biosphere 2 experiment decades ago taught us invaluable, sometimes hard-won, lessons about the delicate ballet of closed environments.

Building these off-world Edens demands a fusion of disciplines – biologist, engineer, chemist, farmer, recycler – all rolled into one. It's about understanding life's intricate web and translating it into reliable technology. The successes harvested on the ISS, the breakthroughs debated in scientific journals, the prototypes humming in labs today – they are the seeds of truly sustainable communities beyond Earth. They are turning the starry-eyed dreams of living on the Moon or Mars into practical blueprints for our future cosmic homesteads.

Forging the Soul for the Silent Sea: How We'll Keep Our Sanity Among the Stars

The starship hums, a tiny bubble of life adrift in the endless, silent dark. Outside, nebulae bloom with impossible colours, but inside, the real frontier isn't the vacuum – it's the landscape of the human mind. Pushing out into the cosmos asks more of us than just engineering marvels; it demands a profound inner strength, a resilience forged specifically for the unique pressures of the Void.

Imagine it: Earth, a beautiful, distant memory. Home is a metal tube shared with a handful of souls. Day and night blur into artificial cycles, sleep becomes a ghost haunted by microgravity, and the silence is only broken by the ship's systems and the static-laced voices from a world light-minutes away. Danger isn't just possible; it's a constant neighbour, whispering in the creaks of the hull and the flicker of warning lights. This cosmic cocktail – isolation, confinement, distance, peril, warped time, relentless work – can corrode the spirit, breeding anxiety, shadows of depression, fractured sleep, and sparks of friction between crewmates who have nowhere else to go. NASA and space agencies worldwide know this: the greatest challenge might not be reaching Mars, but arriving whole.

Mind-Havens in the Machine:

How do you fight the gravity of loneliness? You build anchors for the mind. Researchers are dreaming up lifelines. Picture an astronaut, yearning for sunlight on their skin, slipping on a VR headset. Suddenly, they aren't adrift millions of miles away. They're walking a sun-drenched beach, waves crashing digitally at their feet. They're wandering through a familiar forest, the scent of pine almost real. It's more than escapism; it's mental first-aid. VR is being weaponized against the psychological demons of space – simulating emergencies in safe spaces to build coping muscles, untangling the spatial confusion zero-G can induce, and offering targeted therapy proven effective back on Earth for anxieties and trauma. Think of it as building virtual sanctuaries, pocket dimensions of peace carried across the solar system.

Alongside these digital oases, the old ways still hold power: the rhythm of a strict schedule, the anchor of exercise (a balm for body and soul), sacred windows carved out for sleep, encrypted whispers across the void in private sessions with therapists back home, and the precious, vital lifeline of connection to loved ones – voices and faces crossing the gulf, reminding them they aren't truly alone.

The Shipmate Constellation: Harmony in a Can

In the tight confines of a long voyage, your crew isn't just colleagues; they're your world. The International Space Station hasn't just been a laboratory for science, but a crucible for human connection under pressure. Diaries whispered into recorders, candid debriefings, and careful observations have painted a vivid picture: successful crews aren't just skilled, they click. They build a fragile ecosystem of shared purpose, clear roles, laughter echoing in the modules, and leadership that listens. They know how to talk not just about the mission, but about life, balancing checklists with camaraderie.

We've seen it in the logs: the simple rituals of shared meals, the joy of a makeshift movie night, the collective triumph over a tricky repair, the celebration of birthdays light-years from where they were born – these weave the fabric of morale. But we've also learned from the static, the silences, the documented friction. When personalities clash, when cultural nuances are missed, the thin walls feel even thinner. Conflict resolution isn't a soft skill out there; it's mission-critical survival. The ISS has taught us we need to choose explorers not just for their brilliant minds, but for their resilient hearts and their ability to be part of a fragile human constellation. Performance skyrockets when the crew becomes a true team, a chosen family navigating the dark.

Training Homo Astra: Blueprints for Martian Minds

Now, look further. Imagine not just a mission, but a settlement. A fledgling outpost on the rusty plains of Mars or the stark beauty of the Moon. These pioneers won't be visitors; they'll be the first generation of off-worlders, building lives under an alien sky, potentially forever separated from Earth's embrace. The psychological stakes skyrocket.

How do we prepare them? We draw lessons from Earthly echoes of space – simulated habitats like HI-SEAS perched on volcanic slopes, or the profound isolation of Antarctic winters. These dress rehearsals mimic the confinement, the communication lags, the sheer otherness, letting us watch, learn, and adapt. We see monotony bite, tempers fray over shared resources, and different rhythms clash. And we test the cures: structured downtime, protocols for peace-making, leadership that empowers.

The training ground for these future Martians and Lunarians must be as rigorous for the mind as it is for the body. We're talking:

Forging Inner Shields: Advanced mental resilience bootcamps – mindfulness not just as a practice but a survival tool, cognitive therapies tailored for alien landscapes, stress drills that inoculate against the inevitable shocks.

Weaving the Tribe: Intensive training in the art of "us." Mastering cross-cultural communication for diverse settler teams, learning negotiation when resources are scarce, building systems where pioneers support each other, becoming each other's keepers.

Practicing the Unthinkable in Pixels: VR won't just be for therapy; it'll be the crucible. Trainees will face simulated habitat failures, resource crises, and complex social conflicts, honing their reactions and decision-making under virtual fire.

Living the Future, Now: Long-haul simulations on Earth that aren't just about the hardware, but about replicating the intense social pressures, the isolation, and the psychological marathon of building a society from scratch, far from home.

Ultimately, becoming a multi-planetary species is humanity's boldest leap. But the journey outward demands an equally profound journey inward. Building mental fortitude, fostering deep crew cohesion, and preparing the spirit for the profound challenge of settling new worlds – this is how we ensure that when we plant our flags on distant shores, our minds are as ready as our machines. We are not just exploring space; we are exploring what it means to be human, under the light of alien suns.

Beyond Survival: Crafting Homes in the High Frontier

We talk about leaving Earth, but what if we could take the best parts with us? Not just the people, but the feeling of home? That's the heart beating inside the concept of an O'Neill cylinder. It's less a space station, more a world-ship, an ark carrying not just life, but lifestyle.

The cold equations of space travel tell a harsh truth: our bodies weren't made for perpetual floating. Our bones thin, muscles fade – we're creatures of gravity, tethered to the familiar pull of down. Gerard O'Neill looked at this problem not as a barrier, but as an engineering challenge with an elegant, almost audacious solution: if space won't give us gravity, we'll make our own.

Picture this: vast, sunlit valleys curving gently upwards into an artificial sky. Not a cramped tin can, but enormous cylinders, miles wide and stretching longer still, engaged in a slow, stately waltz through space. This gentle spin isn't just mechanics; it's the foundation of normalcy. It's the force that keeps your feet on the ground, lets you pour a drink without it escaping, allows children to grow strong under a familiar, albeit manufactured, pull. The sheer scale O'Neill envisioned – like his "Island Three" concept – wasn't just for show; it was crucial for comfort, ensuring the spin would be a barely perceptible hum beneath the surface of life, not a disorienting whirl.

Sunlight on Tap and Engineered Edens

Powering these orbiting cities wouldn't be a matter of plugging into the wall. It would require drinking deeply from the sun's firehose. Imagine immense fields of solar panels, glittering wings stretching out to capture the raw, unfiltered energy of space. This isn't just about keeping the lights on; it's about illuminating entire landscapes, driving industries, and sustaining the intricate life support systems that mimic Earth's own breath. It's harnessing a star to build a new kind of world.

Inside? That's where the true magic happens. O'Neill didn't just design a structure; he sketched a landscape. Think rolling hills, forests, rivers, and lakes, all contained within the cylinder's inner curve. Sunlight, gathered by external mirrors, would stream through colossal windows, painting the land with the familiar rhythm of day and night, even the subtle shifts of engineered seasons. This isn't just about illumination; it's about psychological grounding, the deep human need for the sky, for weather (controlled, perhaps, but present), for the horizon.

Urban planning here becomes world-building. Where do people live? In communities that feel familiar, perhaps nestled by an artificial river or overlooking a park that curves overhead. How do they eat? From hyper-efficient farms – vertical gardens reaching towards the central axis, hydroponic bays glistening under grow lights – woven seamlessly into the environment. Industry might hum quietly near the low-gravity center, while transportation, maybe sleek light rail, follows the gentle arc of the ground. Recreation wouldn't be a luxury; it would be essential – hiking trails that loop around the world, lakes for sailing under an artificial sun, spaces designed purely for joy and connection.

This isn't merely about escaping Earth; it's about recreating its essence, perhaps even improving upon it, in the most unlikely of places. It's about building not just habitats, but homes – sustainable, vibrant, and deeply human cities among the stars. The engineering challenge is colossal, yes, bordering on the mythical. But the dream? The dream is profoundly human: to carry the light of community and the feeling of home into the vast, silent darkness.

Powering the Journey: Propulsion Futures

The Burning Ascent: Our Chemical Stairway to Heaven

Imagine humanity, standing at the edge of an infinite ocean – the cosmos. For generations, our vessels for crossing even the nearest shores have been powered by a controlled fury: chemical rockets. Think of it less like a gentle sail and more like strapping ourselves to a barely contained explosion. At its heart is a simple cosmic truth, the one Newton whispered: push something back, and you fly forward. We hurl superheated gas downwards with incredible violence, and in return, the universe shoves us skyward.

How well we do this, how much 'oomph' we get for every precious drop of fuel burned, is measured by something called specific impulse (Isp). It's like the fuel economy of the gods. Our finest machines, gulping down frigid liquid hydrogen and oxygen, boast impressive Isp numbers, touching 465 seconds in the perfect vacuum of space. They are magnificent beasts, the pinnacle of this fiery alchemy.

Physics Slams the Door: The Tyranny of the Rocket Equation

But here's where the dream of interstellar voyages collides with a brick wall named Tsiolkovsky. His famous rocket equation isn't just math; it's a cosmic decree that governs our ambitions. It starkly tells us: want to go really fast? You need either unbelievably efficient engines (way beyond what chemicals offer) or a launch vehicle so gargantuan, so overwhelmingly full of fuel, that the actual payload – the probe, the people – becomes a laughable speck.

The energy locked within chemical bonds, the very stuff that powers our rockets, is finite. It's potent, yes, but fundamentally limited. Even if we built flawless, multi-stage rockets that shed weight like magic, the cold, hard physics suggests we might, at the absolute theoretical limit, nudge a tiny payload towards 2-3% of light speed. To push even a modest probe to a single percent of light speed using chemical rockets? We'd need to launch a fuel tank bigger than our oceans. It's simply not a path to the stars. Chemical rockets are our powerful tugboats, brilliant for leaving Earth's harbor and exploring our local solar system, but interstellar deep-sea exploration demands a completely different kind of ship.

Clever Hybrids and the Air-Breathing Beast: Finding Loopholes

But humanity doesn't give up easily when told "no." Limitations breed genius. Enter the intriguing world of hybrid rockets – a clever cocktail mixing solid fuel (think a sturdy log) with a liquid or gas oxidizer (the 'air' it needs to burn). They promise a simpler, safer approach than purely liquid engines, often allowing us to throttle or even restart the engine mid-flight – a trick most solid rockets can't manage. They have their own quirks, like sometimes burning unevenly, but engineers are relentlessly ironing out the kinks.

Then there's the true wild card, the revolutionary concept engine that sounds like science fiction: SABRE (Synergetic Air-Breathing Rocket Engine). Born from the minds at Reaction Engines Ltd. in the UK, SABRE isn't just hybrid; it's a shapeshifter. Imagine a machine that starts its journey acting like a jet, breathing the very air rushing past it. Only when it's high above the thick atmosphere, screaming along at hypersonic speeds, does it transform, switching to its onboard liquid oxygen tank like a traditional rocket.

The magic enabling this transformation? A piece of engineering sorcery called a precooler. At speeds like Mach 5, incoming air slams into the engine at over 1,000°C – hot enough to melt metal. SABRE's precooler chills this inferno down to a frosty -150°C in less than the blink of an eye (1/20th of a second!), all without turning the air into frost. This lets the engine use the atmosphere's oxygen for the initial, power-hungry climb (up to around 25 km). By "living off the land" for part of its journey, a SABRE-powered craft slashes the colossal amount of heavy oxygen it needs to lug from the ground up. The dream? True Single-Stage-to-Orbit (SSTO) vehicles – spaceships that take off like planes, fly to orbit, and land again, potentially making space access dramatically cheaper and more routine.

Making Today's Giants Dance Better: Honing the Craft

While dream engines like SABRE power towards reality, the relentless drive for better pushes engineers to refine the chemical rockets we rely on right now. This isn't just about raw power; it's about efficiency, cost, and ingenuity:

Stop Throwing Them Away! (Reusability): SpaceX kicked down the door on this one. Their Falcon 9 boosters, returning to Earth hundreds of times (some flying over 20 missions each by early 2025!), proved that reusing the most expensive parts of a rocket drastically cuts the price of admission to space. It shifted the paradigm from disposable machines to workhorses. Starship aims to make the entire vehicle reusable – the next leap in this cost-saving revolution.

Weaving Rockets (Advanced Manufacturing): How we build rockets is changing. Companies like Rocket Lab are 3D printing complex engine parts (their Rutherford engines are full of them!), allowing for faster production and intricate designs impossible before. Relativity Space is even bolder, aiming to 3D print nearly their entire rocket structure. This isn't just printing; it's sculpting lighter, more optimized machines, faster than ever.

Squeezing More Power (Engine Efficiency): Even within chemical rockets, there's room to improve. ULA's Vulcan Centaur, powered by Blue Origin's BE-4 engines, uses a sophisticated "staged combustion" cycle burning methane. Think of it as getting more bang for every buck, extracting more energy from the propellants. Methane is also cleaner-burning (good for reuse) and potentially harvestable on Mars – a fuel for future frontiers.

So, the story of spaceflight unfolds. We are dreamers reaching for the stars, yet tethered by the fundamental laws of physics. But within those tethers, we innovate. We refine. We invent. From the controlled fury of today's chemical giants, honed through reuse and advanced crafting, to the revolutionary promise of air-breathing engines, each step builds on the last. The chemical barrier is formidable, but human ingenuity continues its relentless assault, pushing the boundaries, ever upwards.

Fusion: Stealing Fire from the Sun

Think about it: the Sun, that giant ball of fury hanging reliably in our sky, is powered by nuclear fusion. It's been doing it for billions of years. Our ambition? To build a tiny, controlled star, cram it into the back of a spaceship, and ride its fiery breath across the solar system. Instead of burning refined kerosene, we'd be tricking lightweight atoms – think special kinds of hydrogen (deuterium) and helium – into smashing together under pressures and temperatures that mock earthly comprehension. The sheer energy unleashed would be biblical, heating a gas like hydrogen into a searing plasma and flinging it out the back at speeds that make today's rockets look like they're crawling.

Talk efficiency? Our best chemical rockets are like gas-guzzlers, maxing out around 450-500 seconds of "oomph" (Specific Impulse, or Isp). Fusion promises engines that sip fuel, potentially delivering tens, maybe even hundreds of thousands of seconds of Isp. That's not just an upgrade; it's tearing up the rulebook. It means journeys measured in weeks, not months, and the ability to haul serious cargo – labs, habitats, the works – not just featherweight probes. It's the difference between a weekend trip and an expedition.

DRACO: The Nuclear Engine's First Roar in Space

But bottling a star is hard. Really hard. So, we take steps. Enter DARPA's DRACO project – think of it as the fiery older cousin to fusion, the one paving the way. DRACO isn't fusion; it uses nuclear fission (splitting heavy atoms, like in current power plants) to superheat its propellant. It's still nuclear, still a radical leap. If chemical rockets give us 450 seconds of Isp, DRACO's Nuclear Thermal Rocket (NTR) aims to double that, pushing towards 900 seconds or more.

And it's not just theory! As of now (early 2025), the mission is racing towards a real, live space demonstration planned around 2027. Lockheed Martin is crafting the spacecraft body, while BWX Technologies tackles the nuclear heart. Their goal? To prove this technology works reliably and safely way up there, specifically showing it can zip around the crucial space between Earth and the Moon (cislunar space) with newfound agility.

Why does DRACO matter so much, even if it's not fusion? Because it's forcing us to solve the gnarly problems of handling nuclear power off-planet: mastering reactor control in zero-G, shedding incredible heat in a vacuum, developing materials that don't crumble under radiation, and writing the safety manuals for technologies that demand utmost respect. Every lesson learned with DRACO – every material tested, every control loop perfected – is vital knowledge for the even greater challenge of fusion. Plus, imagine using DRACO-derived tech to cut a Mars trip from a grueling 7-9 months down to 3-4. That's less time soaking in cosmic rays, less strain on human minds and bodies. That makes Mars feel suddenly... closer.

Antimatter: Riding Annihilation (If We Dare)

If fusion is bottling a star, antimatter is like trying to harness the Big Bang itself. Forget fusing atoms; this is about the most violent, efficient energy release physics allows: matter meeting its evil twin, antimatter. Touch them together – electron meets positron, proton meets antiproton – and poof. They vanish, converting their entire mass into pure, raw energy, exactly as Einstein's $E=mc^2$ promised. The power density is simply ludicrous, dwarfing even fusion.

Imagine needing only a pinch – milligrams – of this exotic stuff to fuel interstellar dreams, pushing ships to speeds that could make neighboring stars reachable within a human lifetime. Theoretical designs sketch out magnetic funnels directing the energetic shrapnel of annihilation, or using the blast to superheat propellant or even kick-start a secondary fusion reaction.

But here's the colossal "if." Antimatter is the universe's ultimate diva.

Making it: We can barely create nanograms a year at places like CERN, smashing particles with continent-sized machines. Fueling a starship would require amounts so vast, the energy needed to produce it beggars belief.

Holding it: You can't put antimatter in a bottle; it annihilates anything it touches. It requires sci-fi levels of containment – perfect vacuums, intricate magnetic cages (Penning traps), chilled to near absolute zero – just to keep it from exploding. Keeping that stable on a vibrating, accelerating spaceship? A nightmare.

Using it: How do you safely tap into that annihilation, directing its fury into thrust without instantly vaporizing your own engine?

Antimatter propulsion is the Everest of spaceflight dreams. Overcoming these hurdles would require revolutions in physics and engineering we can barely conceive of today. It's the stuff of distant futures, enabling journeys that are currently pure fiction: zipping across the solar system in days, or sending the first emissaries to Alpha Centauri.

Building the Future, Piece by Fiery Piece

Neither fusion rockets nor antimatter engines will magically appear. The path forward is paved with prototypes, experiments, and probably a fair few explosions in shielded test chambers. DRACO is that crucial prototyping step for nuclear space systems, a tangible bridge from theory to flight.

Elsewhere, scientists wrestle with fusion plasma in giant magnetic donuts (tokamaks) and twisted tunnels (stellarators), primarily for Earth-based power, but pushing the boundaries of materials and control systems vital for space. For antimatter, the "prototypes" are even more fundamental: physicists at places like CERN are fighting to produce a few more anti-atoms, hold them stable for a second longer in complex traps, and run simulations trying to grasp the physics of controlled annihilation.

It's a slow, painstaking process: testing materials until they glow, building smaller test rigs to see if the math holds up, running endless computer models, and finally, integrating everything – reactor, power, nozzle, safety checks – hoping it all works together.

The Journey Ahead

So, that's the roadmap: from the chemical rockets we know, we're stepping up to the nuclear thermal power of DRACO within this decade – a giant leap that promises to shrink our solar system. Beyond that lies the dream of fusion, harnessing the Sun's own power, a goal requiring major breakthroughs but offering staggering rewards. And shimmering on the far horizon, almost mythical, is antimatter – the ultimate power source, guarded by challenges so immense they seem almost insurmountable today.

The work happening now – DRACO's assembly, fusion experiments humming, antimatter particles being trapped – isn't just prep work. It is the journey. It's humanity refusing to be content with the shore, painstakingly building the extraordinary vessels needed to finally sail the cosmic ocean.

Echoes in the Void: Our Reach for Proxima Centauri

Look up tonight. Feel that? That quiet pull of the infinite, the star-dusted velvet stretching beyond comprehension. It's a feeling humans have known forever, the wonder of what's out there. Now, that ancient question is echoing in our labs and engineering bays, shaping into audacious plans. Our nearest stellar neighbor, Proxima Centauri, isn't just a pinpoint of light; it's a destination. But it sits across an ocean of emptiness so vast – 4.24 light-years, a number that translates to trillions of miles and feels like infinity – that reaching it demands we rewrite the rules of travel, perhaps even of being human.

How do we bridge such a soul-crushing distance? Two vastly different dreams are taking shape, born from the same yearning:

1. The Star-Chip Sprinters: Riding Laser Light

Imagine catching sunlight in a sail. Now, imagine that sunlight concentrated billions of times over, focused into a blinding spear of energy fired from Earth. This isn't mythology; it's the heart of concepts like Breakthrough Starshot. We're dreaming of tiny, gram-scale emissaries – think postage stamps packed with the senses of a spacecraft – hitched to sails thinner than whispers, woven from exotic materials yet unborn.

The launchpad? Not a gantry, but potentially square kilometers of Earth's surface transformed into a colossal, ground-based laser array. A "man-made star" designed to unleash a hundred-gigawatt torrent of precisely choreographed light. For a few intense minutes, this beam would slam into the gossamer sail, a photon hurricane accelerating our tiny probe to speeds that defy imagination – maybe 20% the speed of light.

Think about that: 134 million miles per hour. Fast enough to shrink the yawning chasm to Proxima Centauri from a multi-millennia crawl with today's rockets to a mere 20-year dash. It's the difference between a geological age and a human career.

But this breathtaking vision demands miracles:

The Light Fantastic: How do we weave millions of laser beams into one perfectly steady, miles-wide lance of light that stays focused across billions of miles? It's like trying to paint a moving target the size of a dime, four light-years away, using a brush made of atmospheric turbulence and quantum uncertainty.

The Ghost Sail: What can possibly endure that laser's fury, yet be light enough to fly? It needs to be impossibly reflective, strong, stable under immense heat, and tough enough to survive stray interstellar dust motes hitting like microscopic cannonballs. We're talking materials bordering on science fiction.

Worlds on a Wafer: Cramming cameras, navigation, power (maybe sipping energy from isotopes or the laser's glow itself), and a communication laser onto a chip lighter than a butterfly? It pushes nanotechnology to its absolute edge.

Whispers Home: Sending back a picture, even a trickle of data, from 25 trillion miles away with a tiny onboard laser? It requires aiming accuracy that makes threading a needle blindfolded seem trivial, and an Earth-based "ear" vast enough to catch that faint, distant signal.

Starshot is our robotic scout, our first tentative touch. Its goal isn't to carry us, but to carry our gaze – to see if the planet orbiting Proxima Centauri, tantalizingly Earth-sized and in the "just right" zone, holds secrets like air, water, or perhaps even the ghost of life.

2. The World-Seeds: Humanity Adrift for Generations

Then there's the other path. Not a sprint, but an epochal migration. Picture not a tiny chip, but a world set adrift – colossal vessels, kilometers long, carrying not just humans, but a sliver of Earth's biosphere. These are the generational starships, the "slow arks," designed for voyages lasting centuries, maybe millennia.

They wouldn't chase light speed. They'd lumber through the void, perhaps powered by the contained fury of fusion or hypothetical engines scooping interstellar hydrogen (though that idea faces its own giant hurdles). Their challenge isn't sheer speed, but endurance. Absolute, unfaltering endurance.

The logistics are enough to buckle the mind:

Breathing Life into Steel: How do you create a perfectly closed ecosystem – air, water, food – recycling every atom for centuries, supporting thousands of souls within a metal hull? It's building an artificial Earth, stable and self-sustaining, far beyond anything we've ever attempted.

Forever Fix-It: Everything breaks eventually. How do you carry spares for a thousand-year journey, or build advanced factories that can fabricate anything needed from recycled matter, deep in interstellar space? The ship must be its own mechanic, its own mine, its own manufacturer.

The Cosmic Shield: Space isn't empty; it's filled with high-energy galactic cosmic rays, lethal radiation showering down constantly. How do you shield generations of inhabitants without requiring impossibly thick walls? Maybe through massive layers of water or rock, or perhaps even active magnetic force fields.

And beneath the immense mechanics lie questions that pierce our very identity:

The Unchosen Journey: Can someone born on a starship, whose only world is the ship's interior, truly consent to a life dedicated to a mission conceived by ancestors they never knew, bound for a destination they've never seen?

The Soul of the Ship: How does culture evolve, or decay, over centuries of isolation? What kind of governance can possibly remain stable? What happens to the human spirit when "home" is a memory, and the stars outside are a constant, unattainable presence? Could the descendants even remember why they started the journey?

Bridging the Now and the Forever

So, how do we actually plan for Proxima Centauri? These two paths aren't mutually exclusive; they likely represent stages.

Starshot, or something like it, feels like the next step – the daring reconnaissance mission. It's conceivable within our lifetimes or the next, answering the critical question: Is there anything there worth the ultimate journey? It's the robotic eyes we send ahead.

The generational ark? That's a dream for a different era, perhaps centuries from now. It requires not just technological leaps we can barely imagine (like mastering fusion power and true closed ecologies), but also a profound global (or perhaps interplanetary) commitment. It hinges on the findings of those first robotic scouts and a deep, species-wide reckoning with the staggering ethical and existential questions.

Reaching for Proxima Centauri isn't just about building faster ships or better life support. It's about stretching our science, our engineering, our biology, our ethics, and ultimately, our definition of what humanity can achieve. The stars are distant, yes. But the journey – the thinking, the dreaming, the striving – has already begun, whispered on the cosmic wind.

Unchaining Ourselves: Reaching for the Stars Beyond First Gear

Look up. That endless, diamond-dusted blackness calls to something deep within us, doesn't it? It's the ultimate frontier, whispering promises of wonder and discovery. Yet, here we stand, largely gazing from afar. Our trusty chemical rockets, the magnificent beasts that first carried us to the Moon, feel like reliable old steam trains in an age yearning for bullet trains to the cosmos. They're powerful, yes, but they sip fuel measured against the sheer vastness we want to cross. They're bound by the energy locked in their chemical fuels, a fundamental speed limit we're chafing against.

Breaking free isn't just about getting there faster. It's about fundamentally changing the way we explore. Imagine squeezing the Solar System, folding the immense distances between worlds so they become neighbours, not years-long treks apart. This isn't science fantasy anymore; it's the promise shimmering on the horizon, born from bleeding-edge physics now taking shape in engineering labs and sparking to life in early experiments.

From Grueling Marathons to a Martian Sprint

Think about Mars. Right now, strapping humans into a capsule for a trip to the Red Planet is like signing them up for a grueling, six-to-nine-month marathon each way. Picture it: months confined, soaking in cosmic radiation like a sponge, muscles and bones protesting the relentless lack of gravity. It's not just inconvenient; it's dangerous, a heavy tax on the human body. And the stuff! The mountains of food, water, and air needed for that slow boat to Mars bloat the mission's cost and complexity into something astronomical.

Now, let your imagination fly with the engineers dreaming up the next leap. Nuclear Thermal Propulsion (NTP) – think harnessing the contained fury of an atom-splitting reactor to superheat hydrogen into a screaming-hot exhaust – is like shifting into a higher gear. Decades of groundwork (remember NERVA?) and today's powerful simulations suggest NTP could slash that Mars commute in half, maybe down to 3 or 4 months. That's a huge deal. These engines promise nearly double the 'get-up-and-go' (specific impulse, ~900s vs ~450s) of our best chemical rockets. And it's not just theory: NASA and DARPA are building DRACO, aiming to prove an NTP engine works in space by 2027. That's the starting pistol firing on this new race.

But hold on, because the real revolution, the jump that feels like trading the steam train for a starship, whispers from the heart of fusion power. Imagine bottling a miniature star, fusing atoms to unleash colossal energy. Concepts like the Direct Fusion Drive (DFD) cooking at places like the Princeton Plasma Physics Lab aren't just sketching faster engines; they're painting pictures of Mars trips lasting maybe 30 to 45 days. Let that sink in. Mars, not as a distant, years-long commitment, but as a destination reachable in the time it takes for Earth's seasons to shift. The engineering challenges are immense, no doubt, but the physics holds – it's a tangible pathway to turning our solar system neighbourhood into just that: a neighbourhood. Other ideas, like powerful plasma rockets or vast solar sails sipping sunlight, also promise quicker journeys, even if fusion holds the crown for sheer speed potential.

Unlocking the Secrets in the Deep Freeze: Beyond Neptune's Realm

This need for speed isn't just about Mars. Think further out, to the Kuiper Belt – that vast, dark, frozen attic of our solar system beyond Neptune, packed with leftover building blocks from when the planets were born. NASA's New Horizons gave us breathtaking postcards from Pluto and the strange, snowman-shaped Arrokoth, but it took nearly a decade just to get there! Trying to explore that realm properly with today's tech is like trying to map a continent using only stepping stones – achingly slow.

Enter transformative propulsion. Imagine a probe, powered by advanced NTP or maybe even an early fusion drive, blazing across the outer darkness to reach the Kuiper Belt in just a few years. This speed isn't just a convenience; it's a scientific superpower:

Sharper Eyes: Sensitive instruments wouldn't spend years degrading under the harsh bombardment of deep-space radiation. They arrive fresher, ready for their close-ups.

Cosmic Sightseeing: Instead of one fleeting flyby after a decade-long cruise, a fast probe could become a nimble tourist, hopping between multiple, diverse Kuiper Belt Objects, revealing the full story of that mysterious region.

Catching the Action: Imagine spotting something unexpected happening out near Uranus, or even catching an interstellar visitor passing through. Rapid propulsion could mean launching a "rapid response" mission to see it up close before the moment passes.

Science Now, Not Later: Discoveries made years, not decades, from now can fuel the next wave of exploration much faster. It accelerates the entire rhythm of understanding our cosmic backyard.

Picture launching a probe that zips past Neptune in 2-3 years, then spends another few years dancing between icy worlds we've only glimpsed. That's not just a faster mission; it's a fundamentally different, richer kind of exploration, made possible only by tearing up the old speed limits.

Rewriting the Spaceflight Playbook: Agility, Safety, and Ambition

This isn't just about getting places faster; it's about completely changing how we think about and plan space missions.

Bringing Humans Home Sooner: Slashing travel times, especially to Mars, is revolutionary for astronaut health. Less time bathed in radiation, less time weakened by microgravity – it dramatically lowers the risks. Suddenly, maybe we don't need wildly complex, hypothetical spinning spaceships for artificial gravity on those first trips. And think of the packing list! A shorter trip means vastly less food, water, and air to haul from Earth, making the whole endeavor lighter, cheaper, and more achievable.

Unleashing Robotic Explorers: As we saw with the Kuiper Belt, speed means ambition. Mission designers could map out "grand tours" of Jupiter's or Saturn's moons, visiting multiple worlds within a single mission that lasts less than a decade – currently unthinkable. Found something amazing at Enceladus? A follow-up mission wouldn't be 15 years away; it could be planned and launched with newfound agility.

DRACO: The First Real Test: This upcoming mission isn't just about testing an engine; it's about proving we can use NTP to maneuver quickly in space, darting around the Earth-Moon system. Its success will be the first real chapter written in this new playbook, informing how we build faster ships to Mars and beyond.

Dreaming Bigger: The Fusion Explorer: Let's dream for a second. Imagine planning a flagship mission with a fusion drive. It could slingshot past Jupiter, race to Saturn, swing by Neptune, and then spend years touring multiple Kuiper Belt Objects, all within maybe 10-12 years. That kind of comprehensive solar system survey is pure fantasy with today's propulsion; the fuel needs or travel times are just too enormous.

Ultimately, transforming how we push through space transforms our perspective. It changes space travel from slow, lumbering expeditions measured in years or decades, to responsive, agile journeys across the solar system. It makes the cosmos safer for humans, unlocks richer science from our robotic explorers, and crucially, brings the staggering scale of space down to something closer to a human timescale. It's the key to truly becoming a spacefaring people, finally ready to step confidently off our planetary shore.

The Satellite Revolution

Meet the Game Changer: The Humble CubeSat

Picture this: space exploration, but think shoebox, not school bus. That's the heart of the CubeSat revolution. We're talking about incredibly smart little satellites built around a simple, standardized block – just 10 centimeters on each side (like a fancy Rubik's Cube!) and weighing less than a small bag of sugar. This wasn't some corporate masterplan hatched in a sterile lab. Nope, it was born back in '99, cooked up by Professors Jordi Puig-Suari and Bob Twiggs at Cal Poly and Stanford.

Their brilliant idea wasn't about building empires; it was about access. They wanted to give students a real shot at space – a chance to design, build, launch, and talk to their own satellite without needing a nation-state's budget. They wanted hands-on learning, the thrill of creation, the agony of troubleshooting – the real deal. Little did they know, they weren't just building a teaching tool; they were planting a seed that would grow into a forest, disrupting the entire space industry.

The Secret Sauce? Thinking Inside the Box!

What made the CubeSat so ridiculously effective? Standardization. Think of it like Lego bricks for space. By defining that simple 10x10x10 cm "1U" block and its basic rules, Puig-Suari and Twiggs accidentally sparked an ecosystem. Suddenly, companies could design miniature components – radios, cameras, computers, solar panels – knowing they'd fit a standard frame. Deployment systems became standardized too. It was like the PC revolution hitting orbit – suddenly, building a satellite wasn't just for the giant aerospace companies anymore. Costs plummeted, and timelines shrank.

And boy, did the technology catch up! Engineers, fueled by caffeine and ingenuity (and probably presenting their breakthroughs at geek-fests like the IEEE Aerospace Conference), started cramming unbelievable capabilities into these tiny packages. Need to move around in orbit? Miniature propulsion systems, from puffs of cold gas to experimental electric thrusters, gave CubeSats legs. Need to phone home with data? Forget crackly walkie-talkies; sophisticated radios now beam down information at high speeds. Power? Sleeker solar cells and better batteries keep them juiced. Pointing accuracy? Tiny guidance systems now let them aim with precision once reserved for their giant cousins.

From Classroom Dreams to Cosmic Reality

This explosion of accessibility lit a fire under universities worldwide. The original dream of student-built satellites became a vibrant reality. Imagine being a student, not just reading about space, but actually building something that reaches orbit! Think of the University of Michigan's M-Cubed, snapping pictures of Earth, or the massive international QB50 project where dozens of universities teamed up, launching a fleet of CubeSats to study our upper atmosphere together.

These aren't just glorified science projects; they are forging the next generation of space engineers and scientists, giving them cradle-to-grave mission experience that used to be pure fantasy. The numbers don't lie: hundreds of university-led CubeSats have hitched rides to orbit, proving this little box is one of the most powerful educational tools ever launched.

The Power of the Swarm: Thinking Big by Going Small

Beyond the campus, CubeSats rewrote the rules for practical space applications. Why rely on one enormous, eye-wateringly expensive satellite that might take days to revisit the same spot on Earth, and represents a single point of failure? Instead, why not launch a swarm of affordable CubeSats? A constellation working together can blanket the globe, offering persistent, near-real-time coverage. It's strength in numbers, space edition.

And it's happening right now:

Eyes Everywhere (Planet Labs): Planet operates a jaw-dropping fleet of over 200 CubeSats, mostly their "Dove" imagers. They photograph nearly the entire Earth's landmass every single day. This isn't just cool; it's vital for farmers watching crops, environmentalists tracking deforestation, disaster teams assessing damage after a flood or fire, and cities planning their growth. They've even caught illegal mining operations red-handed from orbit!

Weather Whispers & Traffic Cops (Spire Global): Spire's constellation listens to GPS signals bending through the atmosphere to gather crucial weather data (temperature, pressure) worldwide, making forecasts dramatically better. They also act like cosmic traffic controllers, tracking ships (AIS) and planes (ADS-B) across oceans and remote regions, boosting safety and efficiency.

Connecting the Unconnected (Swarm Tech/SpaceX): Got a sensor on a remote farm, a tracker on a shipping container in the middle of nowhere, or a buoy bobbing in the ocean? Swarm's incredibly tiny "SpaceBEE" satellites (a quarter the size of a standard CubeSat!) offer low-cost tendrils of internet connectivity, bringing the Internet of Things to places cell towers can only dream of, for just a few bucks a month.

Digital First Responders: When disaster strikes – an earthquake, a hurricane – the first hours are critical. CubeSat constellations, with their ability to revisit locations frequently, often provide the crucial first images of the damage, helping rescuers know where to focus their efforts, sometimes coordinated through international efforts like the 'Space and Major Disasters' charter.

So, the CubeSat? It's more than just a smaller satellite. It's a democratizer. It's proof that innovation doesn't always need billions of dollars. Fueled by academic curiosity, refined by relentless engineering, and now deployed in powerful swarms, these little boxes didn't just shrink the hardware; they blew the doors wide open on who gets to participate in space and what we can achieve from orbit. They are the embodiment of "Small Satellites, Big Impact."

Our Sky, Awakened: Starlink's Dazzling Gamble

Look up. Just beyond the blue, or veiled by the night, something incredible is happening. Thousands of tiny, sun-catching robots are racing around our planet, a silent, shimmering swarm building a new kind of connection. This is Starlink, SpaceX's audacious dream, not just scribbled on blueprints but actively knitting a digital blanket around Earth. Forget lonely, lumbering satellites parked miles high; we're talking about a bustling city in Low Earth Orbit (LEO), maybe 550 kilometers up – practically skimming the rooftops compared to the old guard. As of early 2025, thousands are already up there, with plans for tens of thousands more, according to the official paperwork.

Why this close orbit? Speed. Pure, exhilarating speed. Being closer slashes the agonizing delay (latency) that plagued satellite internet of the past. Suddenly, video calls from a remote mountain cabin feel seamless, online games respond instantly, and critical data flows like a river, even where fibre optic cables fear to tread. It's more than just adding satellites; it's fundamentally rewiring our planet's nervous system, shrinking distances and potentially dissolving the digital divide, one beam of light at a time. The sheer scale approved by regulators like the FCC signals a seismic shift – we're trading lonely lighthouses in the orbital ocean for a dense, buzzing network right overhead.

The Cosmic Traffic Jam: Dancing with Danger

But imagine launching thousands of new cars onto an already busy highway, where everyone's doing 17,000 miles per hour, and some of the existing traffic is just unpredictable junk. That's the challenge Starlink faces. LEO isn't empty. It's littered with ghosts: dead satellites, old rocket parts, and fragments from past collisions – a minefield of cosmic shrapnel. Experts at places like the European Space Agency use sophisticated models (like MASTER) that paint a stark picture: adding these mega-constellations dramatically spikes the chances of near-misses and outright collisions.

The evidence? Starlink satellites are already playing high-speed dodgeball. While the exact number of swerves is company secret, space-watchers analysing public data suggest these satellites are responsible for a huge chunk of all automated collision avoidance maneuvers in LEO. Each dodge burns precious fuel, potentially cutting a satellite's life short, and complicates the already intricate orbital ballet. The nightmare scenario, whispered about since the 70s, is the Kessler Syndrome: one bad crash creating debris that causes more crashes, triggering a chain reaction that could fence off precious orbital highways for generations. Starlink's bots are smart – designed to dodge autonomously and programmed to deorbit themselves quickly (within 5 years, much better than the old 25-year rule). But with so many players in the game, the statistical odds of something going wrong inevitably creep up.

Keeping the Space Lanes Safe: The New Rules of the Road

This isn't science fiction anymore; it's daily orbital logistics. We urgently need better ways to manage this increasingly crowded space. Think of it as needing both better maps and smarter traffic lights.

Sharper Eyes on the Sky (SSA): We need to see everything. Our current ground-based radars and telescopes (like the US Space Force's network) are getting supercharged by private companies using new tech and clever AI. Seeing smaller, previously invisible bits of debris is crucial – even something pebble-sized hitting at orbital speeds is catastrophic.

Smarter Coordination (STM): With thousands of potential close calls flashing up daily, humans can't handle the load. Companies like SpaceX rely heavily on automated systems to predict threats and trigger dodges. But true safety requires everyone talking to each other – sharing data openly, agreeing on universal "rules of the road" instead of just whispering between operators. We need a global space traffic control system, powered by AI and cooperation.

Calling the Cleanup Crew (ADR): Preventing new junk is vital, but we also need to tackle the mess already up there. Pioneering missions (like ESA's ClearSpace-1 or Astroscale's ELSA-d) are testing ways to grab and remove dead satellites. Think orbital garbage trucks – a crucial service, though making it affordable and scalable is the next giant leap.

Imagine the Scene: A Cosmic Close Call

Suddenly, an alert: A Starlink satellite (let's call her Starlink-A) and a ghost from the Cold War – a chunk of Soviet space junk – are on a collision course. Impact in 12 hours.

The Warning: Data floods in from military and commercial trackers. Algorithms scream "Danger!" based on trajectory wobbles and probability scores.

The Decision: On Earth, and maybe even onboard Starlink-A itself, automated systems calculate the risk is too high. A digital reflex kicks in. A maneuver is plotted – a tiny nudge, a puff from a thruster to change altitude or speed just enough.

The Check: Before firing, the system quickly checks: Does this dodge put Starlink-A on a collision course with something else? The all-clear is given. The plan is logged and shared instantly with the global space-watching community.

The Sidestep: Starlink-A executes the tiny, precise burn. Tracking stations confirm: she's shifted course. The danger has passed.

This silent, high-stakes drama plays out hundreds of times a week, just for Starlink alone. It's a constant, automated dance on the edge of disaster.

Starlink and constellations like it are undeniably changing life on Earth, weaving us closer together. But this revolution rides on a knife's edge. Mastering the art of space traffic management isn't just about protecting expensive hardware; it's about ensuring our pathway to the stars, and the vital connections it provides, remains open and sustainable for the future. We're learning to be responsible tenants of the space above our heads, even as we reach for the stars.

Our Whispering Sky: Listening to Earth's Heartbeat from Above

There's a kind of magic, isn't there, in picturing our home, our Earth, suspended in the quiet dark of space? Seeing the familiar blues and greens swirled with the white brushstrokes of clouds – it pulls at something deep within us. But that view, that god's-eye perspective, is more than just soul-stirring beauty. It's become our planet's EKG, its vital signs monitor, utterly essential for understanding the rhythm and stresses of our world. High overhead, gliding silently, are our technological senses – extensions of our own eyes and ears – constantly listening to the whispers and shouts of the planet. They gather stories told in light and temperature, helping us translate the complex language of our climate and environment. This isn't just data; it's Earth revealing its secrets, and learning to listen is changing everything about how we live here.

GOES-R: The Unblinking Eye on the Storm's Cradle

Imagine holding a cosmic vigil, never looking away. That's the essence of satellites like the GOES-R series (now our faithful GOES-East and GOES-West). Anchored in the sky, they keep a steady, unblinking gaze fixed on our half of the world. For NOAA, these aren't just instruments; they're our sentries against the storm. Their star player, the Advanced Baseline Imager (ABI), is like upgrading from a grainy photograph to an IMAX movie of the atmosphere. NOAA tells us it scans faster, sees sharper, and perceives more 'colors' of light than ever before. This lets meteorologists witness the very birth of thunderstorms, watching weather systems gather strength and fury in breathtaking, real-time detail. It's less like forecasting, more like watching the atmosphere live, catching the subtle cues that precede nature's dramatic outbursts.

Reading the Hurricane's Fever Chart

Nowhere does this vigilance feel more urgent than when the oceans begin to churn, brewing a hurricane. GOES-R watches, day and night, as these behemoths take shape. For the forecasters at NOAA's National Hurricane Center, this constant stream of images is their lifeline. They learn to read the storm's mood in the temperature of its cloud tops, estimating its power when planes can't dare to fly near [Source: NOAA NHC methods]. And with rapid-fire snapshots taken every minute or less, they can spot the tell-tale signs of a storm organizing itself, tightening its core, perhaps preparing to unleash terrifying power [Source: NOAA GOES-R Program Office]. Adding another layer is the Geostationary Lightning Mapper (GLM) – the first of its kind watching from so high. Think of it as sensing the storm's electrical heartbeat. Studies suggest that when lightning flares up deep within a hurricane's core, it can be like a fever spike, signaling intensification ahead [Source: DeMaria et al., 2012]. While predicting a storm's exact fury remains a huge challenge, these detailed portraits are slowly giving us the wisdom to better anticipate their rage [Source: NHC Forecast Verification].

A Recent Memory: Hurricane Ian (2022): As Ian spun towards Florida, gathering terrifying strength, GOES-East didn't blink. Its minute-by-minute imagery showed the storm's structure locking into place. Simultaneously, the GLM detected crackling bursts of lightning near the core – nature's own frantic signals – adding crucial weight to the urgent warnings being issued on the ground [Source: NOAA analyses].

Global Forest Watch: Bearing Witness to the Wounded Green

Beyond the drama of the skies, our space-borne eyes also track the slower, deeper changes happening on Earth's skin – especially the health of our forests. When forests fall silent to the chainsaw, the plough, or the mine, it's not just trees we lose. We fray the intricate web of life, destabilize our climate, and endanger communities. Global Forest Watch (GFW), born from the World Resources Institute, acts as a global neighbourhood watch for our planet's green mantle. Using data mainly from the tireless Landsat (NASA/USGS) and Sentinel (ESA) satellites, GFW employs clever digital 'detectives' to spot where tree cover vanishes, almost as it happens [Source: GFW "About Us" & Methodology]. Building on foundational mapping techniques [Source: Hansen et al., 2013], GFW translates the satellite whispers into maps anyone can understand, sending out alerts like digital smoke signals when forests are disturbed [Source: GFW Platform]. The numbers are sobering – the tropics alone lost an area of forest roughly the size of Greece in 2021, a loss these watchful eyes help us grasp [Source: GFW 2022 Review].

On the Ground: Indonesia's Peat Guardians: In the precious, carbon-rich peatland forests of Indonesia, GFW alerts are more than just data. They are triggers for action. When an alert signals potential illegal clearing – often a precursor to devastating fires – it helps direct rangers and protectors to the scene, sometimes preventing catastrophe and safeguarding these vital ecosystems [Source: WRI Indonesia reports].

Weaving Wisdom from Whispers: Towards Safer Horizons

The true genius of our eyes in the sky emerges when we weave these different threads of information together. It's not just about seeing a storm or a cleared forest; it's about understanding how one impacts the other, how the whole Earth system breathes and reacts.

Sharpening the Storm Warnings: The detailed atmospheric 'weather reports' from GOES-R feed directly into the complex computer simulations that predict a hurricane's path and power. Better ingredients going in mean a more reliable forecast coming out, giving coastal communities precious extra hours or days to prepare [Source: NOAA NWS plans]. The steady improvement in tracking hurricane paths over the years owes a debt to these ever-sharper eyes above.

When Broken Land Bleeds: A forest isn't just trees; it's a sponge and an anchor. Remove it, and the land itself becomes vulnerable. Heavy rains that once soaked gently into the forest floor can now trigger devastating landslides on bare slopes [Source: e.g., Sidle & Ochiai, 2006]. Water rushes off denuded land faster, swelling rivers and causing worse floods downstream [Source: e.g., Bradshaw et al., 2007]. By overlaying GFW's maps of forest loss with rainfall predictions (themselves improved by satellites) and the shape of the land, emergency managers can pinpoint communities living in the shadow of these compounded risks. Warnings become smarter, preparedness more focused.

Tracking Fire and Smoke: GOES-R's ABI doesn't just see clouds; it feels the heat of wildfires, often spotting them before anyone on the ground does. It then tracks the smoke, crucial information for warning people about dangerous air quality [Source: NOAA GOES-R Fire applications]. Combine this with satellite maps showing how dry the vegetation is, and you get a powerful tool for anticipating where the next blaze might erupt.

In essence, our perch high above gives us a dynamic, living portrait of Earth. From the terrifying speed of a hurricane captured by GOES-R to the slow, heartbreaking creep of deforestation revealed by GFW using Landsat and Sentinel, these orbiting sensors gather the intelligence we desperately need. They allow us to move beyond simply reacting to disasters and towards anticipating them. This foresight, born from translating cosmic whispers into human understanding, is fundamental. It's how we learn to live more wisely, build resilience, and protect the irreplaceable lives and places on this beautiful, fragile planet. The view from above isn't just data; it's becoming wisdom, an essential guide for our shared future.

Beyond the Broadcast: Our Skyward Guardians Get Smarter (and Spookier)

Think "satellite," and maybe your mind jumps to grainy TV signals from afar or that little GPS arrow struggling to find you on a map. For decades, they's been reliable, if somewhat mundane, workhorses in the sky. But hold that thought. High above us, a quiet revolution is unfolding. These orbiting machines are evolving, tapping into the weird and wonderful rules of the universe itself to offer us levels of security and precision we once only dreamed of. Forget simple relays; meet the new generation of celestial guardians, exemplified by China's quantum whisperer, Micius, and Europe's sharp-eyed navigator, Galileo.

Micius: Sharing Secrets Written in Quantum Ink

Imagine a secret so profound, it's protected not by complex math problems that future supercomputers might crack, but by the fundamental laws of physics. That's the magic of Quantum Key Distribution (QKD). It's like inventing a cosmic handshake where any eavesdropper, just by trying to listen in, inevitably messes up the handshake, instantly revealing their presence. For ages, this amazing trick was tied down by earthly limits – the delicate quantum signals faded quickly in fiber optic cables. Sending them through the near-perfect vacuum of space, however, was the game-changer.

Enter Micius, China's pioneering satellite launched in 2016. It wasn't just a satellite; it was a daredevil physicist in orbit, designed specifically to test this wild idea. And test it, it did. In a move that echoed across the scientific world, Micius beamed pairs of quantumly linked photons (think spooky twins that mirror each other instantly, no matter the distance) down to two ground stations in China over 1,200 kilometers apart. It worked. The fragile quantum connection held strong, proving space wasn't just viable, but ideal for these delicate transmissions.

But the real showstopper? Micius hosted the world's first intercontinental video call wrapped in quantum security, connecting Beijing and Vienna – 7,600 kilometers of secrets whispered across continents via space. Micius acted like a trusted go-between, creating separate quantum keys with each city and then bridging the gap. While relying on the satellite itself to be trustworthy isn't perfect quantum security, it was a monumental leap, generating secret codes fast enough to lock down highly sensitive data. Micius shouted to the world: quantum-secure communication from space isn't science fiction anymore. It's the first step towards a global network immune to the brute force attacks that loom over our current digital security.

Galileo: Seeing the World with Superhuman Eyes

While Micius focused on how we talk securely, Europe's Galileo constellation tackled where we are, with astonishing clarity. We all know GPS – usually getting us close to our destination. Galileo, designed as a civilian-led global system, already offers a sharper view thanks to better clocks and signals.

But the real magic is Galileo's High Accuracy Service (HAS). Forget "within a few meters." HAS, delivered freely over the internet and directly from space, offers positioning down to 20 centimeters horizontally (that's less than the width of your shoulders!) and 40 centimeters vertically. Suddenly, farmers can seed fields with pinpoint accuracy, self-driving cars can know exactly which lane they're in, drones can navigate complex environments safely, and managing resources becomes incredibly precise. This isn't just an upgrade; it's a new dimension of interaction with the physical world.

And Galileo is security-conscious too. Worried about "spoofing," where fake satellite signals trick your phone or car into thinking it's somewhere else? Galileo's Open Service Navigation Message Authentication (OSNMA) acts like a digital watermark, letting your device verify the signals are genuine. It's an invisible shield against digital ghosts trying to lead you astray. For ultra-critical tasks, there's even the Public Regulated Service (PRS) – an encrypted, robust signal designed to resist jamming and ensure that essential services keep navigating, even under pressure.

Weaving the Threads: A Future Built on Trust and Precision

So, what happens when you combine Micius's quantum secrets with Galileo's pinpoint, trustworthy location? You get the blueprint for truly resilient global satellite networks. Our modern life – from the blink-of-an-eye stock trades needing perfect timing, to controlling power grids, guiding autonomous ships, or coordinating defense operations – depends utterly on these orbital networks.

Imagine:

Unhackable Finance: Galileo's authenticated timing signals preventing market chaos from spoofed timestamps, perhaps backed by quantum-secure links for ultimate transaction integrity.

Ironclad Infrastructure: Sending un-interceptable commands via quantum channels (like those Micius pioneered) to manage remote dams or pipelines, ensuring saboteurs can't secretly tamper with critical controls.

Truly Autonomous Systems: Self-driving cars or delivery drones navigating with Galileo's centimeter-level precision, shielded from hijacking attempts by its anti-spoofing tech, potentially receiving critical software updates via quantum-secured links.

Galileo is laying down the hyper-accurate, authenticated map right now. Micius proved that the ultimate layer of security – quantum communication – is possible via satellite. While a global network whispering quantum secrets is still on the horizon, needing more satellites and faster key-sharing, the path is clear.

These advancements aren't just technical tweaks. They represent a fundamental shift, turning our satellites from simple messengers into sophisticated guardians capable of leveraging the very fabric of physics to build a more secure, precise, and ultimately trustworthy foundation for our increasingly connected world. The sky above isn't just space anymore; it's becoming the silent architect of our future.

Peering into the Universe

Our Cosmic Senses: How We're Learning to See the Universe Anew

For millennia, we've craned our necks, squinting into the velvet darkness, tracing patterns, telling stories, and asking the biggest questions: Where did it all come from? Are we alone? Our eyes, magnificent as they are, only capture the thinnest slice of the universe's spectacle. But our ingenuity knows fewer bounds. We build extensions of our senses, mechanical marvels designed to pierce the cosmic veil, and right now, we're living through a golden age of seeing. Two titans are leading this charge: the James Webb Space Telescope (JWST), our intricate eye far above the world, and the Extremely Large Telescope (ELT), soon to be our giant, watchful presence grounded here on Earth.

JWST: Tuning into the Universe's Hidden Frequencies

Imagine the universe broadcasts on countless channels, but until recently, we were mostly tuned to just one: visible light. JWST is like giving humanity a brand-new set of cosmic ears, exquisitely tuned to the infrared spectrum – the universe's hidden frequencies. Why this channel? Because it carries secrets. Infrared light slides through the dusty, opaque curtains where stars are born and galaxies clash – regions that are blindingly dark to older telescopes like Hubble. More profoundly, it carries the ancient, stretched-out light from the universe's dawn, whispers from the first galaxies flicking into existence after the Big Bang. JWST, with its stunning gold-plated, flower-like mirror (a staggering 6.5 meters across) and instruments like NIRCam and MIRI, acts like a cosmic archaeologist, catching these faint, time-traveling photons.

And oh, the stories it began telling in 2023! It turned its gaze to worlds orbiting other suns – exoplanets. Take K2-18 b, a planet larger than Earth residing in its star's "just right" zone, 120 light-years away. JWST didn't just see it; it tasted its air. By analyzing starlight filtering through K2-18 b's atmosphere, it found the chemical fingerprints of carbon dioxide and methane. But the real jaw-dropper was a hint, a tantalizing whiff, of dimethyl sulfide (DMS). Here on Earth, DMS is overwhelmingly produced by life, particularly ocean plankton. Finding it out there? It's not proof of aliens, not yet. But it's a thunderclap moment – JWST has given us the power to sniff out potential signs of life on worlds light-years away. That's not just better science; it's a paradigm shift in our search for cosmic company.

Then there's WASP-107b, a puffy, hot world unlike anything in our solar system. JWST painted a dynamic portrait of its skies, revealing not just water vapor, but also surprise ingredients like sulfur dioxide and clouds made of fine silicate grains – essentially, sand! This unexpected chemistry hints at fierce stellar radiation driving complex weather patterns, turning these distant dots of light into dynamic, evolving places with their own unique climates. JWST isn't just finding planets; it's revealing their personalities.

ELT: The Giant on the Mountaintop, Ready for its Close-Up

While JWST floats silently in space, a colossal counterpart is taking shape high in Chile's starkly beautiful Atacama Desert. The Extremely Large Telescope (ELT) is exactly what its name promises – a behemoth with a segmented main mirror stretching an almost unbelievable 39 meters across. Imagine a mirror wider than a basketball court, designed to drink in starlight. When it opens its eye later this decade, it will gather exponentially more light than any telescope before it. Coupled with mind-bending adaptive optics – technology that actively cancels out the twinkle caused by Earth's atmosphere – the ELT promises images of staggering sharpness, revealing details we've only dreamed of.

A Cosmic Duet: The Power of Partnership

Here's where the real magic lies: JWST and ELT aren't rivals; they're partners in a cosmic dance. They are designed to work in concert, playing to each other's strengths. JWST, free from atmospheric interference, provides the stable, deep infrared view, perfect for surveys and initial atmospheric chemistry. The ELT, with its sheer light-collecting power and ability to see different wavelengths (including some visible light JWST doesn't), offers unparalleled detail and follow-up potential.

Picture this: JWST flags an intriguing exoplanet like K2-18 b, giving us that first atmospheric sniff. The ELT could then zoom in, using its immense mirror to potentially tease out the planet's mass by watching the star's wobble, probe different atmospheric layers with higher precision, or even, for nearby systems, attempt the holy grail: directly seeing the planet as a separate point of light, distinct from its star's glare. It's a ground-space synergy, a powerful tag team approach where one instrument sets the stage, and the other delivers the close-up, allowing us to tackle cosmic mysteries with a depth neither could achieve alone.

Dreaming Bigger: How Today's Marvels Shape Tomorrow's Vision

This new era isn't just about the discoveries themselves; it's about learning how to see better. The triumphs and challenges of building and operating JWST – aligning those intricate mirror segments, unfolding the massive sunshield, handling the firehose of data – are invaluable lessons. They are the hard-won experience feeding directly into the blueprints for what comes next.

The discoveries, too, refine our aim. Finding potential biosignatures, even tentative ones, with JWST sharpens the focus for future missions, telling us what capabilities are most crucial. The planned synergy between JWST and ELT establishes a powerful precedent: future progress hinges on combining the strengths of space-based and ground-based observatories, working across the electromagnetic spectrum.

This is happening right now with NASA's next great dream: the Habitable Worlds Observatory (HWO). Envisioned as JWST's successor, its grand ambition is to directly photograph Earth-like planets orbiting stars like our Sun and analyze their atmospheres for definitive signs of life. Its design leans heavily on JWST's technological heritage. Crucially, its planners know the ELT will be scanning the skies alongside it, creating a strategic ecosystem of observation. Lessons from JWST's operations, data pipelines, and global collaborations are all being baked into HWO's development.

The Never-Ending Quest

We stand at a thrilling moment. Telescopes like JWST are already redrawing our cosmic maps, turning abstract concepts into tangible chemistry on distant worlds. The imminent arrival of ground-based giants like the ELT promises another wave of breathtaking revelations. Together, they are more than just machines; they are extensions of human curiosity, testament to our relentless drive to understand our place in the vast, astonishing universe. They are how we continue the ancient human tradition of looking up, no longer just squinting, but truly seeing.

Our Cosmic Hunger: Beyond the Looking Glass

There's a primal urge in us, isn't there? To look up, to wonder, to see what lies beyond the familiar blue sky. For ages, our cosmic eyes were crafted from simple glass and polished metal, lenses and mirrors growing ever larger in our quest to gather faint whispers of light from the void. But we hit a wall. Imagine trying to craft a single, perfect window pane the size of a small bus – gravity, physics, and sheer practicality screamed "impossible!" Did we give up? Not a chance. Humanity, faced with a limit, didn't just nudge it; we found clever ways to vault right over it. We learned to build colossal eyes from intricate mosaics and devised cunning techniques to capture the universe's most ferocious whispers – X-rays – fundamentally changing how we gaze into the deep.

Eyes Like Honeycombs: The Genius of Many Parts

Think of the James Webb Space Telescope (JWST). It's not just a telescope; it's a testament to thinking differently. Its magnificent golden eye, stretching over 21 feet wide, isn't one solid piece. Instead, it's a breathtaking honeycomb of 18 hexagonal mirrors, each a marvel of lightweight beryllium polished smoother than imagination and coated in a whisper-thin layer of gold to snag the elusive infrared light from the universe's dawn. Why hexagons? Why pieces? Because fitting a giant, single mirror into a rocket was a non-starter! This segmented design was the key to launching our biggest eye yet into space.

But the real magic happened after launch. Imagine 18 separate mirrors, jostled by liftoff, needing to act as one flawless surface. Tiny motors, like microscopic fingers, nudged and tilted each segment with unimaginable precision – adjustments smaller than a virus – guided by complex light-reading systems. They achieved a collective focus so perfect it defies belief. This wasn't the first time we'd tried this mosaic approach. Giants on Earth, like the twin Keck telescopes in Hawaii, pioneered this path decades ago with their 36-segment eyes. They even face an extra challenge: Earth's own shimmering atmosphere, which blurs the stars. Their solution? Powerful lasers that create artificial "guide stars" high above, allowing computers to instantly flex the mirrors and cancel out the twinkle, giving us crystal-clear views from the ground. This intricate dance of segments and atmospheric correction isn't just clever engineering; it's the only way we know how to build the next generation of terrestrial titans – gargantuan observatories like the 39-meter Extremely Large Telescope and the Thirty Meter Telescope, now rising to probe the cosmos with unprecedented power.

Catching Cosmic Screams: The X-ray Universe

While segmented eyes capture the universe's subtle infrared and visible light, some cosmic events aren't subtle at all. They scream. Exploding stars, matter tortured as it spirals into black holes – these cataclysms unleash torrents of high-energy X-rays. But X-rays are like cosmic bullets; hit a normal mirror head-on, and they punch right through. So, how do you focus a scream?

Enter NASA's Chandra X-ray Observatory, a masterpiece of "catch it sideways" engineering. Its mirrors aren't dishes; they're nested barrels, smoother than anything nature creates. X-rays zipping into the telescope glance off these iridium-coated surfaces at incredibly shallow angles, like stones skipped across a perfectly still lake. This gentle nudge is enough to redirect the X-rays and bring them to a sharp focus. Chandra's vision is astonishingly sharp for an

X-ray eye, letting us witness the inferno around supermassive black holes, like the beast lurking at our own galaxy's heart, Sagittarius A*. We can't see inside the black hole, of course, but Chandra paints vivid pictures of the superheated gas in its final, frantic moments. It helped us understand the colossal jets fired from the M87 black hole (the first one we ever directly imaged!) and maps the ghostly, expanding shrouds of exploded stars, showing us where the elements vital for planets and life are forged and scattered. Other X-ray explorers, like ESA's XMM-Newton and NASA's NuSTAR, add their own unique perspectives, analyzing the energy "colors" of X-rays or tuning into even more powerful emissions, giving us a richer, multi-layered understanding of the universe's most violent dramas.

Building Bigger, Seeing Better: The Ripple Effect

These breakthroughs – building eyes from pieces and learning to gently guide X-rays – aren't just isolated tricks. They're transforming the very blueprint of cosmic exploration.

Scaling the Heavens: Without segmented mirrors, the dream of 30-meter class ground telescopes would still be just a dream. The lessons learned from Keck and JWST are the bedrock upon which the ELT and TMT stand. Bigger eyes mean catching fainter light (seeing further back in time) and discerning finer details (seeing what things are).

Cosmic Teamwork: True understanding comes from seeing the whole picture. The cool gas JWST sees might be violently heated by processes Chandra observes. Combining the infrared view of stars being born with the X-ray view of the winds from massive stars shaping that nursery gives us a holistic understanding impossible with one wavelength alone. Future missions are designed with this synergy in mind – how can different telescopes work together?

Beyond Astronomy: The relentless demand for perfection – aligning mirrors to nanometers, polishing X-ray optics smoother than any pond – pushes the boundaries of what's possible in measurement, materials, cooling technology, and software control. These advances often find unexpected homes, improving everything from making computer chips to medical scanners.

Dreaming Even Bigger: The triumphs of JWST and Chandra ignite our imagination for what's next. Could we build even larger segmented eyes in space? Can we design X-ray telescopes like the proposed Lynx concept, with vision so sharp and sensitive they could witness the birth cries of the very first black holes?

Mastering the art of the segmented eye and the delicate dance of X-ray optics are more than technological feats. They are extensions of our own innate curiosity, powerful tools forged by human ingenuity to peel back the layers of cosmic darkness. Each advancement lets us probe deeper, witness more extreme physics, and ask ever more profound questions about the universe and our place within it. The journey continues, and the view promises to be spectacular.

Whispers from the Dark: Our Search for Cosmic Neighbors

Imagine standing under a vast, velvet sky, scattered with diamond dust. For centuries, that's all it seemed to be – beautiful, distant, and unchanging. But now? We know better. That quiet darkness is actually humming with possibilities, a sprawling cosmic neighborhood potentially filled with countless families of worlds orbiting their own suns. And deep within us, that ancient, perhaps loneliest question echoes: Are we truly alone?

This question isn't just a philosophical musing anymore; it's the fuel for an incredible adventure. We've built extraordinary eyes to peer into the dark, machines designed to catch the faintest hints of these hidden realms. Leading the charge is a tireless scout called TESS (Transiting Exoplanet Survey Satellite), launched by NASA in 2018. Think of TESS as a cosmic census-taker, patiently staring at star after star across almost the entire sky. It's looking for a specific cosmic "wink" – the tiniest, rhythmic dip in a star's brightness that happens when a planet crosses, or transits, in front of it from our point of view. It's like spotting the shadow of a moth flitting across a distant streetlight.

The sheer number of potential worlds TESS has flagged is mind-boggling. We're talking over 5,000 "TESS Objects of Interest," or TOIs. Each one is a tantalizing clue, a maybe-planet waiting for a closer look. This isn't just data; it's a flood of new possibilities, fundamentally changing our map of the local universe. TESS is clever, too. It focuses its gaze on stars that are relatively bright and close to us. Why? Because finding a hint of a planet there makes it much easier for our other telescopes, both on Earth and in space, to follow up, to zoom in and really get to know these potential worlds. TESS essentially draws us a treasure map, highlighting the most promising spots nearby where cosmic gold might be buried.

But finding planets, even Earth-sized ones, is just the first step on a much longer journey. The ultimate prize for many is finding not just rock and gas, but life. This is where the fascinating field of astrobiology steps in – part science, part detective work, part dreaming. Astrobiologists are the ones figuring out what clues to even look for, searching for "biosignatures." Forget little green men waving; think more subtle chemical fingerprints left behind in a planet's atmosphere, hints that living things might be breathing, eating, or just existing there.

One chemical that gets astrobiologists talking is methane (CH_4). Here on Earth, much of the methane in our air is churned out by life, from microbes in swamps to cows in fields. Finding methane floating in the atmosphere of a distant world, maybe alongside clues like oxygen or water vapor, could be an electrifying hint – a possible scent of biology carried on the cosmic wind.

But here's the tricky part, the cosmic puzzle: Nature has its own ways of making methane without any life involved. Volcanoes can burp it out; certain geological processes can create it too. So, just finding methane isn't a slam dunk. Astrobiologists know context is everything. It's like finding a single footprint in the mud – was it a person, an animal, or just an odd pattern left by the rain? They look for atmospheres that seem fundamentally off, out of balance. Imagine finding lots of oxygen and methane existing together – these gases naturally want to destroy each other. Finding both in large amounts suggests something is constantly producing them, tipping the scales. Could that something be life? It's this kind of atmospheric weirdness, this chemical disequilibrium, combined with other clues about the planet itself, that strengthens the case.

This brings us back to TESS's treasure map. Where do we focus our most powerful tools? We prioritize planets hanging out in the "Goldilocks zone" – that sweet spot around a star where it's not blisteringly hot, not freezing cold, but potentially just right for liquid water to pool on the surface. Water, as far as we know, is the key ingredient for life's recipe. TESS has been brilliant at spotting worlds, from Earth-sized pebbles to larger "super-Earths," nestled right in these potentially habitable regions, especially around the small, cool red dwarf stars that pepper our galaxy.

And this is where the real magic of teamwork happens. TESS finds a promising candidate – say, a rocky world in the habitable zone of a nearby star. Often, telescopes on the ground will confirm its existence and weigh it (using a different technique). If it still looks like a winner, it might get bumped to the top of the observation list for our most powerful eye in the sky: the James Webb Space Telescope (JWST). JWST can do something incredible called transmission spectroscopy. As the planet transits its star, JWST analyzes the starlight that filters through the planet's thin atmospheric halo. It's like reading the ingredients label of the planet's air, searching for the tell-tale signs of water, methane, carbon dioxide – those potential whispers of life.

Think of systems like TRAPPIST-1, with its gaggle of Earth-sized worlds. TESS helped pin down their orbits precisely, making it easier for JWST to study them. Or consider LHS 475 b – TESS gave us the first hint, and JWST followed up to peek at its atmosphere (even if, in that case, it didn't find a thick one, which is still valuable information!). Every observation, whether it reveals an atmosphere rich with possibility or confirms a barren rock, helps us refine our search.

So, see TESS as the wide-eyed explorer, scanning the horizon and shouting, "Look over there!" Astrobiology is the wise guide, telling us what subtle signs might mean we've found something extraordinary. And JWST? That's the powerful zoom lens, allowing us to peer into the details, to sniff the air of worlds light-years away. It's a grand, unfolding detective story, driven by human curiosity, piecing together clues from across the cosmos, inching us closer, maybe, just maybe, to discovering we have neighbours after all.

Subtopic 4: Whispers from the Void: Our Quest to Understand a Mysterious Universe

Imagine standing under a perfectly clear night sky, overwhelmed by the sheer scale of it all. Now, picture this: almost everything you can't see, a staggering 95% of the universe, is made of... well, we're not entirely sure. It's a cosmic mystery novel, and humanity has just opened the first chapter. Two of the biggest characters dominating the plot are dark energy, an invisible force pushing everything apart faster and faster, and the enigma of gravity, especially when it flexes its muscles in the most extreme corners of the cosmos.

These aren't just abstract problems for theorists; they're fundamental questions about the reality we inhabit. To grapple with them, we're not just stargazing anymore. We're building extraordinary tools – think cosmic detectives – like sophisticated probes and universe-scale listening devices, to catch clues blowing in from the deepest reaches of space and time.

Peering Through the Cosmic Fog: Meet Euclid, the Dark Universe Mapper

Think about this: roughly 68% of everything out there is this thing called dark energy. It's like an anti-gravity ghost, embedded in the fabric of spacetime itself, constantly whispering "expand, expand!" We first got a solid hint of its existence when astronomers noticed distant exploding stars seemed dimmer, farther away than expected – the universe wasn't just growing, it was accelerating. But what is this stuff? Is it just a background hum leftover from the Big Bang (Einstein called it a 'cosmological constant'), or is it something more dynamic, a clue that our understanding of physics needs a major rewrite?

Enter Euclid, our eye in the sky launched by the European Space Agency in 2023. Euclid isn't designed to see dark energy directly – you can't! Instead, it's a cosmic cartographer, tasked with mapping how the universe's structure has evolved over the last 10 billion years. It uses two ingenious techniques:

Cosmic Funhouse Mirrors (Weak Gravitational Lensing): Imagine light from a faraway galaxy traveling towards us for billions of years. As it passes massive objects (galaxies, clumps of invisible dark matter), gravity bends its path slightly, like looking through subtly warped glass. Euclid will measure the shapes of billions of galaxies, detecting these tiny distortions. By analyzing how shapes are warped across the sky, we can map out the invisible scaffolding of dark matter and see how the push-and-pull between gravity (clumping things) and dark energy (spreading them out) has shaped the universe we see.

Echoes of the Big Bang (Baryon Acoustic Oscillations - BAO): Right after the Big Bang, the universe was a hot soup of particles and radiation. Sound waves rippled through it. When the universe cooled and atoms formed, these ripples essentially froze, leaving a faint pattern – a preferred distance between clumps of matter, like ripples spreading on a pond. This distance acts as a cosmic "yardstick." Euclid will measure the positions of hundreds of millions of galaxies to see how large this yardstick appears at different times in cosmic history. This tells us precisely how fast the universe was expanding at different epochs, giving us a detailed timeline of dark energy's influence.

Euclid's colossal survey, covering a huge chunk of the sky, will give us the sharpest picture yet of dark energy's behaviour. Is it constant, or does it change? The answer could either confirm our current models or throw open the doors to entirely new physics.

Listening to Spacetime's Drumbeat: LIGO and the Symphony of Collisions

While Euclid maps the grand, quiet structures, another set of instruments is listening for the universe's most violent events. LIGO (in the US) and its partners Virgo (Italy) and KAGRA (Japan) are essentially giant ears designed to hear ripples in spacetime itself – gravitational waves. Predicted by Einstein, these waves are generated when colossal objects like black holes or super-dense neutron stars spiral into each other and merge.

The first time we "heard" such an event, GW150914 in 2015, it was revolutionary. Two black holes, heavier than we thought possible, had danced themselves into oblivion, shaking the universe. It was the dawn of a new kind of astronomy. Since then, nearly 100 such cosmic cataclysms have been detected, each one a unique signal from the edge of known physics.

These aren't just blips in the data; they're incredible opportunities:

Putting Einstein to the Ultimate Test: The exact shape of a gravitational wave signal encodes the details of the merger. By comparing the signals we detect to the predictions of Einstein's General Relativity, we can test his theory under conditions of unimaginable gravity – far stronger than anything in our solar system. So far, Einstein's century-old theory is holding up beautifully.

Hearing and Seeing the Universe (Multi-Messenger Astronomy): In 2017, something amazing happened. LIGO/Virgo detected GW170817, the merger of two neutron stars. But this time, telescopes around the world saw the aftermath – from gamma-ray bursts to a faint glow. This single event was a treasure trove: it confirmed these mergers forge heavy elements like gold, gave us a totally new way to measure the universe's expansion rate, and revealed secrets about the bizarre matter inside neutron stars.

Where the Grand Meets the Extreme: Solving Cosmic Puzzles Together

This is where it gets really exciting. Euclid mapping the cosmic web tests gravity and dark energy on the grandest scales. Is the universe unfolding as expected? LIGO/Virgo/KAGRA test gravity in the most intense environments imaginable. Do the laws hold up when spacetime is bent to the extreme?

And sometimes, they work together on specific puzzles. Take GW170817 again. By seeing both the gravitational waves and the light arrive at almost the exact same instant after traveling for 130 million years, scientists confirmed that gravity travels at the speed of light with incredible precision. This single measurement instantly ruled out many alternative theories of gravity!

The Current Cliffhanger: The Hubble Tension

Right now, there's a nagging cosmic disagreement. Measure how fast the universe is expanding today using nearby supernovae (related to how dark energy was found), and you get one number (around 73 km/s/Mpc). Predict the expansion rate based on the early universe (using data like the cosmic microwave background, similar physics to Euclid's BAO), and you get a different number (around 67 km/s/Mpc). This difference, the "Hubble Tension," is statistically significant and hints that our standard picture of the cosmos might be missing something.

This is where our new tools become crucial arbitrators. Gravitational wave events like GW170817 offer an independent way to measure the expansion rate (using them as "standard sirens"). Future detections will refine this. Meanwhile, Euclid's hyper-precise measurements of the BAO "yardstick" will give us the best-ever reading from the large-scale structure side. Resolving this tension could force us to rethink dark energy, dark matter, or even gravity itself.

In essence, we're living in a golden age of cosmic exploration. With instruments like Euclid and LIGO, we're no longer just passive observers. We're actively interrogating the universe, listening to its whispers, mapping its hidden structures, and testing the very laws that govern reality. The mysteries are deep, but our resolve – and our tools – are sharper than ever. The quest continues.

United in the Stars

Forging Frontiers: Lessons Scrawled in Stardust from Apollo 11 and the ISS

There's a pull, isn't there? A deep-seated human itch to look up, wonder, and then go. Exploring the vast, silent dark isn't just a matter of nuts, bolts, and rocket fuel; it's a reflection of our own complex spirit. It tests our ingenuity, yes, but it also tests our ability to get along, to share a dream bigger than any single nation. Gazing back at two of our most audacious adventures – the defiant leap of Apollo 11 and the intricate orbital dance of the International Space Station (ISS) – reveals fascinating truths about ourselves and offers a compass for the voyages yet to come.

The Handshake Across Orbit: Building the ISS

Imagine this: the Cold War's icy grip finally loosening, and the very rivals who once raced to conquer space decide to build something together. That's the improbable genesis of the International Space Station. It wasn't just ambitious; it was revolutionary. NASA (USA), Roscosmos (Russia), ESA (Europe), JAXA (Japan), and CSA (Canada) – giants of space exploration – pooled their smarts, their spacecraft, and frankly, staggering amounts of money. Pinning down the exact cost is like trying to weigh a cloud, shifting with every calculation of launches, decades of operations, and global research grants. But figures often orbiting north of $150 billion give you a sense of the sheer, unwavering commitment this global consortium poured into their shared outpost among the stars.

But the ISS is far more than the most expensive clubhouse ever built. It's a humming laboratory unlike any on Earth, floating in the strange realm of microgravity. Here, scientists unravel the secrets of how our bodies cope (or don't) with long stretches in space – vital clues for protecting pioneers heading to Mars, learning how bones thin and muscles waste away without Earth's constant pull. They forge new materials, watch over our home planet, and test technologies that will power the next generation of exploration. Just keeping the lights on aboard the ISS is a daily miracle of international coordination, a constant dialogue between control centers oceans apart, blending different engineering philosophies and languages. It's a testament to the fact that even when tensions flare on Earth, the shared goal of science can sometimes carve out a space for remarkable cooperation. This very model of intricate agreements and shared operations is the bedrock upon which new initiatives, like the Artemis Accords aiming for a return to the Moon, are being built.

One Giant Leap, Watched 'Round the World: The Echo of Apollo 11

Now, rewind to July 20, 1969. The world felt different. Locked in the Cold War's fierce embrace, the USA and Soviet Union weren't just competing; they were in a cosmic duel fought with ingenuity and national pride. Apollo 11 was undeniably born from this rivalry. Yet, when Neil Armstrong's boot touched the lunar dust, something shifted. An estimated 650 million souls across the globe – a staggering portion of humanity then – were glued to flickering screens, watching not just an American triumph, but a human one. For a breathtaking moment, borders seemed to blur, replaced by a shared gasp of awe.

Though driven by a desire to be first, Apollo 11 became an unintended gift to the world. It ignited a passion for science and technology that rippled across continents, inspiring countless young minds to dream of stars and circuits. The sheer technological might required to reach the Moon spun off innovations that touched everyday life back on Earth – from stronger materials and better computing to medical breakthroughs and even safer food. While a victory claimed under one flag, the images beamed back – a fragile blue marble hanging in the black – spoke a universal language of vulnerability and wonder. Apollo 11 proved that even a fiercely competitive leap can, paradoxically, foster a profound sense of shared destiny.

Blueprints for Tomorrow's Frontiers: Lessons Learned

So, what do these two epic sagas – one a global embrace, the other a national sprint heard 'round the world – teach us as we plot journeys to the Moon, Mars, and beyond?

Build on Trust, Seal it with Ink (ISS Lesson): The ISS didn't just happen; it was meticulously planned. Complex international agreements laid out who did what, who paid for what, and how decisions were made. Future grand projects, like those envisioned under Artemis, need this same bedrock of clear, legally sound frameworks from the start. Goodwill is lovely, but shared blueprints prevent cosmic arguments.

Weaving Ambition with Alliance (Apollo & ISS Lesson): Apollo proved that national drive can achieve the seemingly impossible. The ISS shows that sustained, complex progress often demands shared shoulders. The trick for future missions is finding that sweet spot: letting partners pursue their own goals (developing tech, national pride) while contributing to a larger, shared adventure. It's the art of the cosmic "win-win."

Weathering Earthly Storms (ISS Lesson): Space cooperation isn't immune to politics back home, but the ISS has shown remarkable resilience. Keeping the station flying high, even when international relations frayed, took deliberate effort and dedicated communication channels. Future partnerships need to anticipate political turbulence and build in buffers and reliable ways to talk, compartmentalizing the mission from disputes below.

Lighting the Fire of Imagination (Apollo & ISS Lesson): Both missions, in their own ways, captivated humanity. Apollo's sheer audacity, the ISS's quiet, persistent presence – they remind us why we explore. Future ventures must actively share their stories, their discoveries, their "wow" moments. Public excitement isn't just nice; it's the fuel that ensures long-term support back on Earth. Remember the global thrill of Apollo – that's the kind of inspiration we need to keep fostering.

Speaking the Same Engineering Language (ISS Lesson): Getting hardware built by different teams in different countries to literally plug-and-play in orbit is a monumental challenge. Finding common ground on things like docking systems, or enabling technology sharing where possible (navigating tricky export rules and secrets), makes everyone's life easier and cheaper. The ISS forced engineers to create clever ways to connect disparate systems – a vital skill for assembling future bases on the Moon or Mars piece by international piece.

In the end, our cosmic journey, vividly illustrated by the shared triumph of the ISS and the world-uniting moment of Apollo 11, tells a clear story. Competition can light a fire under us, pushing innovation at breakneck speed. But the truly grand, enduring endeavors – the ones that build lasting footholds off-world – increasingly demand that we join hands. By absorbing the lessons etched in the history of these missions – the financial muscle required, the scientific treasures gained, the operational ballets performed, and the sheer human impact – we can design our next giant leaps more wisely. The stars beckon, and experience teaches us we reach them faster, and perhaps more meaningfully, when we map the journey together.

Beyond the Blue Marble: Handshakes and Moonshots

Imagine stepping onto the dusty plains of the Moon, not as a lone explorer, but as part of a bustling new frontier town. Like any pioneers setting up camp, you'd need some ground rules, right? A way to make sure everyone plays nice, shares the sandbox, and doesn't accidentally trip over each other's moon buggies. That's the heart and soul behind the Artemis Accords.

Think of them less as a stuffy legal treaty and more as a global handshake agreement, spearheaded by NASA but embraced by a growing family of nations (over 30 and counting!). Rooted in the granddaddy of space rules, the 1967 Outer Space Treaty, these Accords are basically a promise: "Let's explore space peacefully, openly, and safely, so everyone benefits." It's about establishing cosmic etiquette before the neighborhood gets too crowded.

What kind of promises are we talking about? Stuff that just makes sense:

Play Fair: Use space only for peaceful adventures, no cosmic squabbles.

Show Your Work: Be open about what you're doing up there and share the cool science you find. Trust is built on transparency!

Build Bridges, Not Walls: Design spacecraft and systems that can actually work together. Imagine needing a rescue and having incompatible docking ports!

Lend a Hand: If an astronaut's in trouble, everyone helps. It's the spacefarer's code.

Keep Track: Let everyone know what objects you've put into orbit so we don't have celestial fender-benders.

Respect History: Those iconic Apollo footprints? Yeah, let's not bulldoze over them. Protect our shared space heritage.

Mine Responsibly: If we find valuable stuff on the Moon or asteroids, let's figure out how to use it sustainably, following the rules we already have.

Give Each Other Space: Coordinate missions to avoid stepping on toes or interfering with experiments.

Clean Up After Ourselves: Mitigate space junk! Nobody wants a cluttered cosmic backyard.

Scientists constantly crunching numbers on space traffic jams and orbiting debris tell us these kinds of agreements aren't just nice-to-haves; they're essential if we want to keep space usable and safe for generations to come. The Accords are the scaffolding upon which we're building our shared future beyond Earth.

India's Lunar Dream: Dusting Off the Moon's South Pole

While global handshakes set the stage, individual nations are still writing their own epic poems in the sky. Take India's Chandrayaan saga, a journey powered by the ingenuity of the Indian Space Research Organisation (ISRO). It's a tale of ambition, learning, and ultimate triumph.

Chandrayaan-1 (2008): The trailblazer! This orbiter didn't just map the Moon; it delivered the stunning news – there's water ice up there! Confirmed with help from international partners (like NASA's instrument riding shotgun), this changed everything we thought we knew about lunar resources.

Chandrayaan-2 (2019): Upping the ante with an orbiter, lander (Vikram), and rover (Pragyan). The final moments of the landing were heartbreaking as Vikram crash-landed.

But failure? Far from it. The orbiter soldiered on, sending back priceless data, and the hard-won engineering lessons paved the way for what came next. Resilience in action.

Chandrayaan-3 (2023): Redemption! Learning from the past, ISRO nailed it. On August 23rd, 2023, the world watched as the Vikram lander gently touched down near the Moon's mysterious south pole – a place no one had ever landed before. India became only the fourth nation to soft-land on the Moon and the first to conquer the challenging southern region. The little rover, Pragyan, rolled out, tasted the lunar soil, and beamed back its findings. Crucially, just months before this historic landing, India signed the Artemis Accords, signaling its commitment to teamwork even as it flexed its incredible national capability.

Cosmic Barn-Raising: How It All Fits Together

So, how do these global agreements and national triumphs actually mesh? They allow us to dream bigger, together. The Artemis Accords, with their focus on things like playing well together (interoperability) and sharing knowledge, provide the common ground.

Look at the Lunar Gateway, the planned space station orbiting the Moon. It's NASA's brainchild, sure, but it's being built like a magnificent international LEGO set. NASA's handling the main living quarters and power, Europe's adding more living space and refueling tech, Canada's contributing a super-smart robotic arm (Canadarm3!), and Japan's chipping in with logistics and habitat parts. It's a cosmic potluck where everyone brings their best dish!

This kind of complex, multi-nation venture only works because of pre-agreed standards and a shared rulebook – exactly what the Accords foster. Building giant projects in space is mind-bogglingly expensive and risky. Teaming up spreads the cost, shares the technological hurdles, brings together the brightest minds from around the world, boosts the scientific discoveries, and hey, it's pretty good diplomacy too!

Ultimately, our next giant leaps won't be taken just by one nation planting a flag. They'll be powered by this beautiful blend: the audacity of national dreams like Chandrayaan, guided by the wisdom of shared principles like the Artemis Accords. It's this synergy – individual brilliance woven into a tapestry of cooperation – that will truly carry humanity into the final frontier.

Our Cosmic Crossroads: Charting a Shared Destiny Beyond Earth

Imagine standing at the edge of the greatest ocean ever known – not water, but the infinite, starry expanse. For generations, we dipped our toes in, planting flags and sending probes like messages in bottles. But now? Humanity's cosmic neighborhood is bustling. Low Earth orbit buzzes like a city square with satellite constellations, dreams of lunar villages echo in mission control rooms, and the lure of celestial resources whispers of both untold bounty and potential conflict.

This isn't just about rockets and robots anymore; it's about us. As we step more boldly off our home planet, we face a profound choice: Do we carry our old terrestrial squabbles and short-sighted habits into the heavens, cluttering the void and drawing new battle lines? Or can we, finally, look up together and forge a truly unified path forward? The future of humanity beyond Earth hinges on crafting a shared dream, building bridges between nations in space, and writing the rules for this incredible new chapter before the story writes itself in ways we regret.

A Global Handshake in Orbit: Why We Need a UN Space Guild

Think of our current space rules, like the venerable Outer Space Treaty of '67. It was a beautiful, hopeful promise – space belongs to everyone, no WMDs allowed! But it was drafted when only a couple of giants walked the cosmic stage. Today, nearly half the world's nations have satellites, and private companies are launching thousands more, turning orbital lanes into crowded highways. The old rulebook feels like using horse-and-buggy laws for drone superhighways.

This is where the idea of a dedicated Space Council, perhaps nestled within the embrace of the United Nations, starts to feel less like bureaucracy and more like common sense. We need more than just a registry of who launched what (though UNOOSA does vital work!). We need a global forum with real teeth – a place to act as cosmic traffic control, mediate disputes before they escalate, set binding agreements on how to handle the growing storm of space debris, and ensure the riches of space aren't just a grab-fest for the powerful few. Imagine a "mission control" for humanity's shared spacefaring future, built on the UN's foundation of global legitimacy. Getting there won't be easy; it demands patience, diplomacy, and learning from past attempts at global cooperation. But it's the necessary next step to prevent the final frontier from becoming another fragmented, contested territory.

Cosmic Dodgeball: The Unifying Threat That Makes Us Look Up

If anything can remind us we're all adrift on the same fragile "pale blue dot," it's the shared vulnerability to cosmic impacts. Asteroids – remnants from the solar system's birth – aren't just sci-fi plot devices. Remember Chelyabinsk in 2013? A relatively small rock, exploding in the sky, injuring hundreds and rattling a city. It was a stark reminder that the universe occasionally throws things our way, and borders are meaningless to a speeding space rock.

This existential threat demands we work together. It's a planetary defense drill we can't afford to fail:

The Cosmic Watch: We need eyes everywhere – a global network of telescopes on Earth and vigilant sentinels in space (like the upcoming NEO Surveyor) scanning the celestial neighborhood for potential threats. Sharing this data instantly, through networks like the IAWN, is like a global neighborhood watch for incoming danger. We've found many big ones, but countless smaller, city-threatening rocks are still out there.

Know Thy Enemy: Spotting isn't enough. We need to understand these celestial wanderers – their size, spin, makeup – to know how to deal with them.

The Nudge: Actually deflecting an asteroid is the ultimate team sport. NASA's DART mission was groundbreaking – humanity literally nudged an asteroid! But other methods might be needed, requiring global R&D efforts, perhaps coordinated by groups like SMPAG under the UN's watchful eye.

No single nation has the resources, the technology, or the geographic coverage to handle this alone. Planetary defense is the ultimate "all hands on deck" scenario, a powerful reason to strengthen global cooperation, perhaps under the guidance of that proposed UN Space Council. It's about protecting our shared home, together.

Writing the Rules of the Road for the Final Frontier

So, what does this look like on the ground (or rather, in orbit)? It means rolling up our sleeves and drafting the practical, everyday rules for living and working in space, guided by that unified vision:

Orbital Traffic Jams: The skies are filling with swarms of satellites. It's amazing, but it's also treacherous. We desperately need internationally agreed-upon rules for who flies where, how to dodge collisions, sharing location data, and crucially, how to responsibly remove satellites when they die. Right now, it's mostly voluntary guidelines – like suggesting drivers stay in their lanes without actually painting the lines or having traffic lights. We need binding agreements before a cascade of collisions (the dreaded Kessler Syndrome) traps us beneath a shell of our own junk.

Cleaning Up Our Cosmic Backyard: Decades of launches have left ghosts in orbit – dead satellites, rocket stages, fragments from collisions. This space junk threatens working satellites and future missions. We need policies that assign responsibility and create incentives (or mandates) for developing and deploying technologies to actively remove the most dangerous debris. Think cosmic sanitation engineers, working under global guidelines.

The Lunar & Asteroid Resource Question: As missions like Artemis aim for the Moon and companies dream of mining asteroids, we're facing a new gold rush. The Outer Space Treaty says nations can't claim territory, but what about pulling resources out? It's a grey area wide enough to drive a spaceship through. Initiatives like the Artemis Accords are one approach, but we need a truly global conversation, likely under the UN umbrella, to establish fair rules for extraction, sharing the benefits, and protecting these celestial environments before conflicts erupt over who owns a patch of lunar ice or a mineral-rich asteroid.

Crafting these policies requires a potent mix of diplomacy, science, and law. A dedicated UN Space Council could be the crucible where these elements fuse, turning the collaborative spirit forged in facing shared threats (like asteroids) into tangible rules for a sustainable and peaceful future in space. It's about moving from simply reaching for the stars to becoming responsible stewards of the cosmos, ensuring this vast frontier remains an arena of opportunity and wonder for all humankind, for generations to come.

The Final Frontier as the First Bridge: Finding Common Ground in the Cosmos

Look up tonight. Beyond the city lights, past the familiar moon, lies an infinite expanse – cold, silent, challenging. We call it space. We send rockets piercing through the blue, deploy satellites like watchful eyes, and dream of footprints on other worlds. But space is more than just a destination for our machines and our scientific curiosity. It's becoming something unexpected: a vast, neutral ground, a silent theater where the often-fractured family of nations can practice the art of getting along.

Forget embassy halls and tense negotiation tables for a moment. Imagine diplomacy conducted not with stern faces across polished wood, but through shared blueprints for a satellite, a helping hand offered across continents via orbital data, or even a literal handshake in zero gravity. This is space diplomacy – using the shared challenge and potential of the cosmos to weave threads of cooperation, offer aid that truly uplifts, and build goodwill that resonates back down to Earth.

A Hand Up From Orbit: Sharing the Stars' Bounty

Think about the power held within a single satellite image. For a nation vulnerable to floods or droughts, it's not just data; it's foresight, a chance to prepare, a map for rescuers. For a farmer struggling with unpredictable weather, it's a digital lifeline, helping manage precious water or optimize harvests. This is the heart of space aid. Organizations like the UN's UNOOSA champion the idea that the tools we use to explore the stars can radically change lives on Earth, especially in developing nations. Programs like UN-SPIDER act as translators, turning complex satellite information into actionable knowledge for disaster response – literally using eyes in the sky to keep feet safe on the ground.

It's often not about grand gestures like donating entire satellites (though that happens!). More often, it's about sharing the know-how: training local scientists, collaborating on tools tailored to specific needs – perhaps monitoring precious forests or managing growing cities. It's about giving access to the invisible networks that guide our world, like the GPS, Galileo, or BeiDou systems that underpin everything from efficient shipping to smarter farming, unlocking enormous economic potential.

Consider India's South Asia Satellite: Launched in 2017, it was presented as a "gift" to its neighbors, a shared resource for communication, weather forecasting, and disaster warnings. A gesture aimed at knitting a region closer together through shared technology.

Or China's Space Silk Road: An ambitious plan weaving space cooperation – navigation services, satellite launches, ground stations – into its broader Belt and Road Initiative, using cosmic infrastructure to deepen earthly connections and influence.

These aren't just acts of charity; they are investments in partnership, subtle ways for nations to demonstrate leadership not through power, but through shared progress and responsibility.

Orbital Handshakes: Finding Unity in the Void

Nothing quite captures the essence of space diplomacy like seeing astronauts from different nations – perhaps even rival nations – working shoulder-to-shoulder in the fragile bubble of the International Space Station (ISS). Born from the thaw of the Cold War, the ISS is a marvel not just of engineering, but of sustained international trust. The US and Russia, alongside partners from Europe, Japan, and Canada, built and operate this outpost together.

Its very existence demands cooperation. Keeping the lights on, the air breathable, the station orbiting requires constant communication, shared protocols, and a deep, unavoidable reliance on each other. Even when political storms rage on Earth, the necessity of collaboration aboard the ISS often keeps vital channels open between technical and governmental communities. It's a powerful, practical peace-building exercise playing out 250 miles above our heads.

Think back to 1975: the Apollo-Soyuz Test Project. An American Apollo and a Soviet Soyuz spacecraft docking in orbit. A symbolic handshake high above the Iron Curtain, proving that even in the depths of geopolitical winter, common ground could be found by reaching for the stars. The ISS is the enduring legacy of that moment. The astronauts themselves, sharing meals, challenges, and the profound "overview effect" – seeing Earth as a single, borderless home – become powerful ambassadors for shared humanity.

The Ripple Effect: Cultivating Goodwill from Above

Ultimately, both targeted space aid and grand cooperative missions achieve something vital: they build goodwill. Space exploration captures the imagination like little else. Those breathtaking images from Webb, the sight of an international crew floating together, the knowledge that satellite data helped avert a crisis – these stories resonate globally. They boost a nation's image, demonstrating technological prowess and a commitment to peaceful, shared endeavors. This is 'soft power' in its purest form – winning hearts and minds not through coercion, but through inspiration and shared benefit.

When nations work together in space, they send a powerful signal. The European Space Agency (ESA) embodies this – multiple nations pooling resources to achieve goals none could alone, like the Copernicus Earth observation system or Galileo navigation. And crucially, ESA often makes its data freely available worldwide, a gesture of cosmic generosity that builds trust and empowers others.

Space, then, is far more than a void. It's a stage where we can act out our better angels. By sharing the tools of space to empower those most in need, by embarking on complex missions that demand we trust one another, and by simply sharing the awe-inspiring journey of exploration, we find unique ways to connect. Reaching for the stars, it turns out, might be one of the best ways to bring us closer together right here on Earth. As we venture further out, the importance of this cosmic diplomacy will only continue to shine brighter.

Laws of the Final Frontier

Stardust Dreams and Earthly Rules: Our Great Leap Upward

Look up tonight. Really look. That vast, inky canvas, scattered with diamond chips and the Moon's gentle gaze, isn't just emptiness anymore. It's humming. It's the new Wild West, the new Silk Road, the ultimate high ground, all rolled into one. And it's pulling at something ancient within us – that itch to explore, to plant a flag, to lead the pack, and maybe, just maybe, to strike it rich among the stars. Launching skyward isn't just poetry now; it's a planet-wide obsession, a potent cocktail of national pride that puffs chests, science that pushes boundaries, strategy played out on a cosmic chessboard, and the undeniable lure of fortunes waiting to be made. As giants like America and China fix their eyes back on the Moon and dream of Martian footprints, the paths they carve reveal fascinating truths about who they are and how they imagine their destiny under that endless sky.

America's Artemis: A Cosmic Potluck with Audited Recipes

Picture NASA's Artemis program not just as building a moon base, but like organizing a massive, slightly chaotic, interstellar potluck dinner with ambitions to eventually cater Mars. This isn't one nation conquering the void; it's shaped by the noisy, passionate debate of a town square – specifically, the US Congress. They hold the wallet, deciding whether the next billion goes to the giant SLS rocket (the main course transporter?) or the Orion capsule (the emergency ride home?), making sure the folks paying the taxes feel like they have a seat at the planning table. And hovering nearby are the meticulous auditors, like the GAO, peering over the cosmic grocery list, asking sharp questions: "Are we sure we can afford this decades-long feast? Is the kitchen being run efficiently?" Their reports are the dose of reality, the reminder that even star-bound dreams need solid ground beneath their launchpads.

But the real flavour of Artemis? It's the open invitation. The Artemis Accords are like passing around a sign-up sheet for galactic good manners, based on old space rules but aimed squarely at the future. By early 2025, over forty nations had added their names, promising to play nice, be open about their plans, lend a hand in emergencies, and share the cool rocks and data they find. Sure, it's partly about sharing the astronomical costs and effort – space is hard! – but it's also America trying to gently suggest the rules of the road, especially for sticky issues like tapping into lunar water ice or minerals. Think of it as proposing the etiquette for this new frontier. And America isn't just inviting other countries; it's hiring pioneers. Through programs like CLPS, NASA's basically paying private companies millions to act as cosmic delivery drivers, sending robotic scouts and supply runs to the Moon. It's a gamble on grassroots ingenuity, hoping to spark a bustling off-world economy, drive down launch costs, and make lunar visits feel less like epic sagas and more like regular commutes – weaving the hustle of private enterprise right into the fabric of its national space dream.

China's Tiangong: The Dragon's Lair Among the Stars

China's ascent, perfectly captured by the gleaming Tiangong space station orbiting proudly since late 2022, sings a different tune. It's a soaring anthem of national triumph, built on fierce self-reliance and executed with pinpoint precision. Run by the CNSA, Tiangong is presented to the world as China's stunning celestial laboratory, a powerful symbol of its technological prowess and undeniable global influence. Official statements often wrap it in the language of national strength and visionary destiny, a message broadcast with confidence.

Often perceived as charting a more solitary course – partly thanks to US laws that slam the door on direct NASA-CNSA partnership – China hasn't locked itself away entirely. Instead, it has carefully laid out its own welcome mat. The CNSA speaks of peaceful exploration and shared benefits, but collaboration happens on China's terms: equality, mutual respect, and progress defined together. The partnership with the UN's space office (UNOOSA) is the perfect example: hand-picking experiments from various countries to fly on Tiangong. It's an open door, yes, but China definitely holds the key, managing access through clear, state-approved procedures ("Measures for Management..."). This ensures any international involvement aligns neatly with China's own strategic roadmap and operational control. It's a masterful way to build diplomatic bridges and showcase Tiangong's capabilities, all while protecting national interests and demonstrating its distinct approach to space diplomacy – an invitation to a meticulously curated gallery, rather than an open house.

The Cosmic Dance: My Star, Your Star, Our Sky?

So, here we stand, humans peering into the infinite, wrestling with a timeless dilemma played out against a backdrop of stars: How do we chase our own national ambitions in a realm that, by its very nature, feels like it should belong to everyone, yet currently belongs to no one?

Moon Dust and Diplomacy: Think about lunar resources – that potential water ice or those rare minerals. The Artemis Accords basically say, "Let's agree on some ground rules based on old treaties and start figuring out how to use this stuff," which helps clear the path for future moon mining and bases. But others, notably Russia and China, lean towards hashing out brand-new, globally agreed-upon rules within the broader United Nations framework before anyone starts digging. It's the classic tension: act now based on your group's understanding, or wait for the whole world to agree? How we solve the "who gets to mine the Moon and how" puzzle will be a massive test of balancing national hunger with global fairness.

Orbital VIP Lounges vs. Open Doors: Consider the homes we build in orbit. For years, the International Space Station (ISS) was the symbol of a broad, if sometimes strained, partnership. Among the partners, access was relatively open. Tiangong, while welcoming specific international guests (like those UNOOSA experiments), feels more like an exclusive club, curated and controlled by China – a sovereign choice reflecting a desire for order and national oversight. Meanwhile, US law (the Wolf Amendment) acts like a strict bouncer, explicitly barring China from the ISS party, showcasing how national security fears can draw hard lines even in the vacuum of space. Who gets invited where, and who is pointedly left off the guest list, paints a stark picture of how earthly politics are carving up the heavens.

Ultimately, both the United States and China are using the vast, dark canvas of space to paint stunning self-portraits of their national hopes and styles. America weaves a complex pattern of alliances and enterprise, checked and balanced by its democratic heartbeat. China constructs a formidable edifice of self-sufficiency, engaging with the world through carefully controlled gateways. The next great chapter of human history, the one written beyond Earth, depends entirely on how these giants, and every other spacefaring nation, learn the delicate ballet between reaching for their own star and remembering that the infinite darkness connecting us all requires bridges, not just higher walls. Finding that harmony isn't just about crafting clever policy; it's the profoundly human challenge that will shape our destiny among the stars.

Strapping In for the Stars: Navigating the Rules, Risks, and Rocket Dreams of the New Space Age

We're living in it – the era where reaching for space isn't just for superpowers anymore. It's the age of "NewSpace," a buzzing ecosystem of private dreamers and doers aiming for the cosmos, launching satellites, planning orbital factories, maybe even hotels with an incredible view. This isn't your grandparents' space race; it's a faster, nimbler, and wildly ambitious surge. But like any frontier, this rush needs some rules of the road, and that's where things get interesting.

The FAA: Trying to Conduct a Rocket Symphony

Imagine trying to choreograph a ballet where every dancer has invented their own style, built their own shoes, and wants to leap right now. That's roughly the challenge facing the Federal Aviation Administration (FAA) in the U.S. They're the ones tasked with keeping this explosive growth from literally exploding unsafely.

Their big move? Part 450. Think of it as throwing out the old, rigid instruction manuals (Parts 415, 417, and the rest) and replacing them with something more like: "Show us you know how to drive safely, no matter what kind of car you've built." Finalized in late 2020, this was meant to be a breath of fresh air, letting companies innovate how they meet safety goals, rather than just ticking boxes. Got a revolutionary rocket design? Great! Prove it's safe your way. One license could even cover multiple flights from different spots for up to five years – flexibility was the name of the game.

But reality bites. The sheer speed of NewSpace – with launch numbers jumping over 30% last year alone (148 licensed operations!) – is straining the system. Innovators and even politicians are tapping their feet, wondering if the "streamlined" process is getting bogged down. It's like paving a highway while rush hour traffic is already zooming past. Recognizing this, the FAA hit the 'pause-and-reflect' button, gathering experts in late 2024 (the SpARC committee) to figure out how to smooth the bumps in Part 450, particularly around proving flight safety. Congress even called in the watchdogs (the GAO) to see if the rules are truly working as intended. It's a living system, constantly tweaking the dials to match the industry's velocity.

The Billion-Dollar Question: Who Pays if Things Go Wrong?

Space is big, but accidents can have down-to-Earth consequences. This is where decades-old international agreements step in, forming a kind of global safety net. The granddaddy, the 1967 Outer Space Treaty, basically says: your country is responsible for everything launched from its soil or by its companies. Launch something? You own the consequences.

The 1972 Liability Convention gets specific. If your rocket debris hits someone's house or a plane? Your country pays, no ifs, ands, or buts (that's "absolute liability"). If your satellite bumps into another satellite way up there? Liability depends on who was actually at fault. It's a system designed to ensure someone, ultimately a nation, foots the bill.

How does this work day-to-day? Insurance. The FAA makes launch companies get hefty third-party liability insurance. This brings in the specialized world of space insurance – underwriters betting on incredibly complex machines braving extreme environments. They worry about new tech they haven't seen before, the increasingly crowded highways of orbit (especially with tiny satellites often flying uninsured), and finding the sweet spot between "enough coverage" and "bankrupting the startup." This structure, where nations bear ultimate responsibility backed by private insurance, is crucial. Imagine telling a brand-new rocket company they face unlimited liability for a catastrophic failure – innovation might just freeze on the launchpad.

The Startup Gauntlet: Earning Your Ticket to Ride

For the passionate teams in garages and small hangars dreaming of orbit, getting that FAA Part 450 license is the mountain to climb. It's not just paperwork; it's proving your rocket science is sound before you even get to the main application. You need to convince the FAA you've thought through every possible way a flight could go wrong, how you'll handle toxic materials, and that your critical safety systems are rock-solid. We're talking detailed maps of where debris might fall, deep analysis of potential hazards, and blueprints of your tech.

The experience on the ground varies wildly. You hear stories like William Bruey's from Varda Space, who felt tangled in red tape for reentry, lamenting that regulations seemed to be holding back humanity's reach. It's the kind of frustration that fuels calls for reform. Yet, others, like Justin Fiaschetti at Inversion, found the process manageable, stressing that talking to the FAA early and doing your homework makes all the difference. Even giants like SpaceX operate under Part 450 for their cutting-edge Starship tests, showing the rules apply to iterative rocket science, though often accompanied by intense public debate over environmental impact.

The Balancing Act: Fueling Dreams, Ensuring Safety

So, governing the private space boom is like trying to safely harness lightning. Part 450 is the FAA's attempt at a modern, adaptable harness, but the sheer energy of the industry is testing its limits. International treaties provide the ultimate backstop, while insurance markets handle the immediate financial shocks. And for the startups carrying the fire of innovation, navigating the path to launch requires technical brilliance, deep pockets, and patient persistence.

As humanity pushes further and faster into the final frontier, driven by both profit and pioneering spirit, the rulebooks, the risk calculations, and the very definition of "possible" will keep evolving. It's a high-stakes, high-speed journey, and we're all watching, wondering what's next.

Our Footprints on Forever: Finding Our Way Among the Stars

We stand at the edge of the infinite, humanity blinking in the starlight, ready to take tentative steps beyond our blue cradle. Probes whisper secrets from distant planets, and dreams of footprints on the Moon and Mars feel less like science fiction and more like destiny knocking. But as we pack our cosmic bags, the cargo isn't just technology and ambition; it's our conscience. We're not just launching rockets; we're launching ourselves – our values, our flaws, our hopes – into the vast, silent unknown. How do we walk softly on worlds that aren't ours? How do we ensure this grand adventure isn't just another story of the powerful claiming the prize, but a truly human saga? Charting our course requires more than just star maps; it demands a moral compass calibrated for the cosmos, one that guides us through the fog of planetary protection, the thorny path of fairness, and the practical challenge of writing rules for realms we've barely touched.

Whispers of Life, Echoes of Responsibility: The Sacred Silence of Other Worlds

Imagine standing on Mars. The silence is profound, broken only by the wind. Is this silence empty, or does it hold the sleeping secrets of life, past or even present? This is the heart of our first great ethical test: planetary protection. It's a cosmic version of "leave no trace," magnified a billionfold. What if our arrival, our microscopic hitchhikers from Earth, becomes an invasive wave, silencing alien whispers before we even learn their language? It would be a scientific heartbreak, an irreversible act of cosmic vandalism.

Conversely, what if we bring something back? An unknown microbe, an extraterrestrial 'Andromeda Strain,' unleashed unknowingly into our own vibrant, fragile biosphere? This tightrope walk between exploration and contamination is guided by scientists at COSPAR, who act like cautious librarians of the solar system. They categorize worlds and missions, setting strict "cleanliness" standards [Source: COSPAR Planetary Protection Policy documents]. Think of Mars' "Special Regions" – potential oases where Earth life might just thrive – as the most restricted areas, demanding almost ritualistic purification of our robotic envoys [Source: COSPAR Mars Special Regions definition]. We bake spacecraft, bathe them in potent vapors, all striving for near-impossible sterility [Source: NASA Planetary Protection Procedures]. Why? Because hardy Earth bugs have proven they can survive the void and Martian-like hellscapes [Source: Astrobiology journal studies], making the threat real. We're trying to keep the odds of accidentally seeding a Martian water source lower than winning a lottery – less than 1 in 10,000 [Source: COSPAR quantitative guidelines].

It's not theoretical. NASA's rovers, like Perseverance, are assembled in hyper-sterile chambers, technicians gowned like surgeons attending a birth [Source: NASA mission documentation]. Yet, is it enough? Or is it too much, strangling exploration in costly red tape? And who holds the fast-growing private space ventures to these same sacred vows? The ghost of Israel's Beresheet probe, spilling its cargo of near-indestructible tardigrades onto the Moon, reminds us that our cosmic footprints can be accidental, indelible marks [Source: Beresheet mission reports].

Sharing the Starlight: Is Space for Everyone, or Just the Elite?

The 1967 Outer Space Treaty sings a beautiful song: space is for all humankind, a shared inheritance [Source: Outer Space Treaty, Article I]. But the reality hums a different tune, one scored by exorbitant price tags and technological hurdles. Launching dreams costs billions – look at the magnificent James Webb Space Telescope, a $10 billion eye on the universe [Source: NASA JWST Fact Sheet]. This staggering cost creates a celestial divide. While some nations build starships, others struggle to access the vital Earth-watching data satellites provide – crucial eyes in the sky for farming, disaster relief, or tracking climate change [Source: UN reports on Space for Sustainable Development]. It's the digital divide stretching to orbit.

Are we creating a future where the benefits of space – knowledge, resources, inspiration – flow only to those who can afford the ticket? Initiatives like the UN's Access to

Space for All try to level the field, offering rideshares for satellites or experiments [Source: UNOOSA initiative descriptions], but it often feels like handing out rowing boats while others command fleets.

Consider the vast constellations of satellites, like Starlink, promising global connection. A noble goal, certainly. But astronomers squint, their view increasingly marred by streaks of artificial light. Experts worry about a crowded sky turning into a demolition derby (the dreaded Kessler syndrome), and latecomers finding the best orbital highways already choked with traffic, owned by private gatekeepers [Source: Astronomical Union statements, space debris research]. Sharing the stars means more than just planting flags; it means ensuring the pathways, the views, and the discoveries are genuinely open.

From Noble Ideas to Star Laws: Crafting the Rules of the Road

So, we have principles – protect other worlds, share the journey. But principles are whispers until they become policy, agreements with teeth. COSPAR's guidelines are respected suggestions, not interstellar law. The Outer Space Treaty is a foundational constitution, but it lacks the chapters needed for today's bustling, commercializing space age.

We need to translate our moral compass into practical governance. This isn't just bureaucratic shuffling; it's about building the framework for a peaceful, productive future off-Earth. How do we ensure everyone, including private companies, adheres to planetary protection? How do we create cosmic traffic cops and rules of the road to prevent disastrous pile-ups in orbit? When we start mining the Moon or asteroids, who owns the treasure, and how do we share the wealth, honoring that "province of all mankind" promise? How do we ensure the firehose of data pouring back from space becomes a global library, not a private collection?

Initiatives like the US-led Artemis Accords attempt to sketch some answers, creating guidelines for lunar neighbors on transparency, safety, and resource use [Source: NASA Artemis Accords documentation]. But they also spark debate: is this the best way forward, or does it risk creating exclusive clubs, bypassing the slower, more inclusive UN process where all voices can be heard [Source: Space policy journal analyses]? Finding the right path requires balancing speed with consensus.

Our Cosmic Character

Pushing into the cosmos is more than reaching new destinations; it's about deciding who we become when we get there. Ignoring the ethical shadows means risking a future marred by contamination, conflict, and cosmic inequality – repeating the darkest chapters of exploration on Earth, just on a grander stage. Protecting alien quiet, ensuring the ladder to space has rungs for everyone, and building fair rules of the road – these are the pillars of responsible exploration. By listening to science (like COSPAR), honoring foundational ideals (like the Outer Space Treaty), and forging practical agreements together, we can strive to make our journey to the stars a testament to the best of humanity. The universe is vast, mirroring not just our ambition, but the weight of our shared duty. The stars are watching. Let's show them our best selves.

Healing Among the Stars

Our Bodies, Whispering Betrayals in the Starry Void

Imagine it: letting go, truly floating free. No up, no down, just effortless existence among the stars. It's the dream of space, right? But while our minds might soar, our very bones begin a quiet rebellion, missing the familiar anchor of Earthly gravity. Our skeleton isn't just scaffolding; it's a living, breathing construction site, constantly tearing down and rebuilding itself. Gravity is the foreman, the constant, gentle pressure that shouts, "Stay strong! We need you!"

Take away that foreman, float into the microgravity sea, and the worksite goes haywire. The demolition crew – those little cells called osteoclasts that nibble away old bone – keeps punching the clock, maybe even works a bit of overtime. But the builders – the osteoblasts tasked with laying down fresh, strong bone – they mostly shrug, put their feet up, and wait for a load that never comes. The signal to build is gone, replaced by the silent void.

The result? It's like a slow-motion heist happening inside our own bodies. Our bones start shedding their density, their very essence, at a startling rate. Forget the slow creep of aging on Earth; up there, astronauts can lose 1% to 2% of bone mass in crucial spots like hips and spines every single month. Picture cramming a decade of Earthly bone thinning into a single six-month tour on the International Space Station. That's the invisible price of floating free. The groundbreaking NASA Twins Study, comparing astronaut Scott Kelly's year in orbit to his Earth-bound twin Mark, painted a vivid picture of how space fundamentally rewrites our biology, reinforcing this stark reality about our skeletal foundations.

The Fight Back: Wrestling Gravity Where There Is None

Losing bone isn't just a number on a chart. It means returning to Earth – or stepping onto Mars – with bones that have become terrifyingly brittle, like fine china where steel is needed. A simple stumble could mean a devastating fracture. And there's a nasty side effect: all that calcium leached from the bones floods the bloodstream, potentially clogging the kidneys and forming painful stones.

So, how do we protect the explorers charting these new frontiers? We fight back. We make gravity happen. On the ISS, astronauts aren't just lounging; they're locked in a daily, grueling 2.5-hour battle against this decay. They wrestle with machines like ARED (Advanced Resistive Exercise Device), using powerful vacuum cylinders to mimic the heavy lifting – the squats, the deadlifts – that tells their bones, "We still need you strong!" They pound on treadmills, strapped down by harnesses to simulate the impact of running under gravity.

It's a relentless effort, supported by careful diets rich in calcium and Vitamin D, though pills alone can't trick bones into staying dense. Scientists are even exploring bone-boosting drugs used for osteoporosis on Earth, but the space environment adds its own complex twists. For those epic voyages to Mars, lasting years, perfecting this internal defense is non-negotiable. We need astronauts to arrive, explore, and return home not as fragile shells, but as strong pioneers.

Cosmic Storms and Invisible Bullets: The Radiation Siege

Beyond the body's internal struggles lies an external threat: the universe itself. Here on Earth, we're nestled under a cozy blanket – our atmosphere and magnetic field – that deflects the worst of space radiation. Venture beyond, towards the Moon or Mars, and you step into a cosmic shooting gallery.

Two main culprits lurk out there. First, the Galactic Cosmic Rays (GCRs): tiny, high-energy bullets – fragments of atoms stripped bare – fired from exploded stars light-years away. They are relentless, coming from all directions, punching through spacecraft and flesh like microscopic shrapnel. They shred DNA, planting seeds for cancer, clouding the eyes with cataracts, and potentially damaging the brain over the long haul.

Then there are the Solar Particle Events (SPEs): sudden, violent outbursts from our own Sun. Imagine a solar flare unleashing a flash flood of energetic particles. While GCRs are a chronic threat, a big SPE can deliver a massive radiation dose in hours, posing an immediate, acute danger if astronauts aren't protected.

Building Havens in the Storm: Shields Against the Invisible

How do you build a shield against something so pervasive and energetic? It's trickier than you might think. Intuitively, you'd pile on lead, right? Great for X-rays here. But against GCRs, heavy materials can be disastrous. Hit a lead atom with a high-energy cosmic ray, and it can shatter, creating a spray of secondary particles that might be even worse than the original bullet.

The surprising hero? Humble hydrogen. Materials packed with hydrogen – think plastics like polyethylene, or even good old water (H_2O) – are much better "catchers' mitts." The light hydrogen nuclei are good at absorbing the energy of incoming particles without creating that dangerous secondary shrapnel. It's about cleverly slowing down and breaking up the cosmic bullets.

So, future spacecraft, like the Orion capsule destined for deep space, are designed with "storm shelters." Not bunkers of lead, but cleverly arranged areas packed with water tanks, supplies, and blocks of hydrogen-rich plastic, creating safe havens for the crew during intense solar storms. For future Martian or Lunar bases, the dream is to use the local dirt – the regolith – perhaps mixed with water or hydrogen-rich binders, to build thick, protective habitats. It's a constant balancing act: creating effective shields without weighing the spacecraft down so much it can barely leave Earth.

Protecting humans from themselves (their own bodily responses) and from the universe (its invisible radiation) – that's the profound challenge, and the incredible adventure, of reaching for the stars.

Stitching Life Lines Across the Stars: Healthcare Beyond Earth's Embrace

The dream glitters – humanity, standing on Mars, watching Earth rise over a lunar horizon. It's a breathtaking vision. But beneath the stardust, a fragile truth remains: we are human. Vulnerable. How do we mend a broken bone, soothe a fever, or face a medical emergency when home is a distant blue marble, days, weeks, or even months away? Earth's hospitals, with their familiar hum and vast resources, become impossible luxuries. Out there, in the quiet, demanding vastness, healthcare must be reborn – tougher, smarter, and woven into the very fabric of the journey. This isn't just about packing better first-aid kits; it's about forging a new kind of medicine, ready for the final frontier.

Echoes in the Void: Ultrasound as Our Eyes Across the Emptiness

Picture this: An astronaut, floating within the metallic womb of the International Space Station (ISS), doubles over. A sharp, insistent pain flares deep inside. Panic is a luxury they can't afford, but the nearest emergency room is 250 miles straight down, through layers of atmosphere. Giant MRI machines? Power-hungry CT scanners? They're back on Earth, laughably impractical for space.

But there's hope, nestled amongst the station's complex machinery: ultrasound. It's compact, gentle (no harsh radiation), a window into the body. Yet, the true magic isn't just the device, but the connection it enables. For years, the challenge was clear: you need an expert to wield the probe, to interpret the subtle dance of shadows and echoes. How do you do that when the experts are grounded?

The answer whispers across the void: telemedicine, reimagined. Imagine a doctor on Earth, thousands of miles away, becoming the eyes and hands of a crewmate onboard. Through crystal-clear video and calm, precise instructions, they guide the astronaut's hand holding the ultrasound probe. It's a digital handshake across the cosmos. "Move it slightly left... angle it down... hold it there." Suddenly, Earthbound eyes see inside the astronaut's body, diagnosing kidney stones, checking strained eyes adjusting to microgravity's pull (that tricky SANS condition), even watching the heart beat and fluids shift in this strange new environment. They can hunt for dangerous blood clots (DVT), a hidden risk of floating free from gravity's pull. This isn't just remote assistance; it's a vital, near-real-time diagnostic lifeline, proving its worth time and again, making ultrasound an indispensable tool in the deep-space medicine chest.

Weaving Hope from Within: Bioprinting Skin on the Frontier

Out there, the body faces new betrayals. Wounds seem to mend reluctantly, stitch by painful stitch, potentially slower than on Earth. And the risks? A flash burn from equipment, the invisible sting of radiation beyond our planet's magnetic shield – trauma is a constant companion. Packing enough spare skin grafts for every 'what if' on a three-year trip to Mars? The sheer bulk and weight make it impossible.

This is where the future unfolds, cell by living cell. At the Wake Forest Institute for Regenerative Medicine (WFIRM), pioneers are crafting a near-miraculous technology: printing skin. Not synthetic sheets, but living tissue, tailored for the patient. Imagine taking a tiny sample of an astronaut's own healthy skin cells. These cells are nurtured, multiplied, then mixed into a special "bio-ink." Then, like an artist painting on a canvas, a highly specialized printer lays down this cellular ink, layer by delicate layer, directly onto the wound – a burn, a deep gash.

The beauty of this for space travelers is profound:

Healing On Demand: Forget bulky supplies. Print what you need, when you need it, using the astronaut's own biological toolkit.

Perfect Harmony: Because it's their own cells, the body welcomes the graft, minimizing the risk of rejection – a critical advantage when isolated.

A Potential Healing Boost: By placing the right cells exactly where they're needed, this method promises to coax the body's healing processes, perhaps even overcoming the sluggishness seen in space.

While still being perfected on Earth, the dream is clear: An astronaut, seared by Martian radiation during a surface exploration, wouldn't just be bandaged and hope for the best. A small sample, a few days or weeks of cellular expansion onboard, and then – the quiet hum of the bioprinter, weaving a living bandage directly onto their injury. It's regenerative medicine, meeting the harsh realities of space head-on.

The Cosmic Clinic: Building Resilience into the Walls

Designing a sick bay for the Moon or Mars isn't just about cramming Earth-tech into smaller boxes. It's about building a sanctuary of healing that's as tough and independent as the pioneers it serves. Distance changes everything. A 40-minute round trip radio signal to Mars means "real-time" surgical guidance is impossible. This extraterrestrial clinic must be:

Ingeniously Compact: Every ounce, every watt of power, is precious. Think multi-tasking marvels, instruments shrunk down yet powerful, seamlessly integrated.

Indestructibly Robust: It has to survive the bone-jarring violence of launch, function flawlessly in strange gravity, and shrug off cosmic radiation, all while needing minimal hands-on fixing.

Intelligently Autonomous: With Earth potentially silenced by distance, the clinic needs its own brain. Artificial intelligence must become the co-pilot, analyzing scans, spotting danger signs in vital data, offering treatment pathways, maybe even guiding robotic hands through delicate tasks. Imagine AI whispering guidance through an augmented reality headset to a crew medical officer facing their first major emergency.

Frugally Self-Sufficient: Water, air, resources – everything must be conserved, recycled, intertwined with the ship's life support. Even medicines need rethinking – how do they last for years? Perhaps tiny labs will compound drugs on demand.

Think of it: An astronaut on Mars develops searing abdominal pain. Appendicitis? The delay to Earth is too long for interactive surgery. Instead, the onboard AI analyzes the ultrasound (guided by a crewmate using stored protocols), reads the blood test from a lab-on-a-chip, checks the vitals. Diagnosis confirmed. The AI then projects step-by-step surgical instructions, perhaps onto smart glasses, guiding the designated medical officer, or even assists a nimble robotic arm. This isn't science fiction; it's the necessary evolution for survival far from home.

Ultimately, reaching for the stars demands we become masters of healing in isolation. The tele-guided ultrasound, the promise of bioprinted regeneration, the intelligent, self-reliant clinic – these aren't just tools. They are extensions of our human determination to care for one another, no matter the distance. They are the stitches binding our fragile biology to our boundless ambition. And remarkably, the solutions forged in the crucible of space often find their way back, improving medicine for all of us right here, under the familiar blue sky of Earth. The voyage outward demands innovation inward, pushing the frontiers of medicine as surely as we push the frontiers of space.

The Silent Ocean Within: Charting the Mind's Voyage Through Deep Space

We stand on the precipice, dreaming of crimson Martian plains and the stark beauty of lunar horizons. The call of the cosmos is intoxicating, a siren song pulling humanity towards the unknown. Yet, as our machines grow ever more capable of bridging interplanetary gulfs, we confront a profound truth: the longest, most treacherous journey isn't measured in miles, but within the fragile confines of the human mind. Sending people into the deep black for months, or years, means confronting the shadows that lurk in isolation – particularly the heavy cloak of depression.

Imagine a world shrunk to the size of a small habitat, perched on the desolate, volcanic slopes of Mauna Loa. This is HI-SEAS, Earth's meticulous, sometimes brutal, dress rehearsal for life far from home. Here, brave crews willingly step into the silence, the confinement, the profound aloneness of simulated space missions. What echoes back from these enclosed worlds? The undeniable strain on the human spirit.

Think of it: the same few faces, day after relentless day. The unchanging walls pressing in, a constant reminder of the infinite emptiness outside and the shrinking personal space within. News from loved ones arrives across a ghostly time lag, amplifying the already crushing distance. The world outside your window is magnificent, perhaps, but utterly alien and unchanging. It's a recipe for the mind to turn inward, sometimes towards darker shores.

HI-SEAS doesn't just hint at this; it paints a clear picture. Researchers track the subtle shifts in mood, the creeping fatigue, the friction that inevitably sparks between crewmates breathing the same recycled air for too long. They've seen morale dip sharply mid-mission – a kind of psychological deep winter known as the "third-quarter phenomenon." It's not just feelings; sometimes, the body betrays the stress, releasing hormones like cortisol as if facing a tangible threat. The threat, however, is the suffocating monotony, the relentless pressure to perform, the ache of separation. These simulations confirm: the isolation needed for space travel can inadvertently cultivate the seeds of depression.

But humanity is not adrift without an anchor. Just as we engineer marvels of technology, we also possess ancient, powerful tools for navigating our inner landscapes. Enter meditation – not as a vague wish for calm, but as a scientifically validated practice for forging mental resilience.

Consider Mindfulness-Based Stress Reduction (MBSR). It teaches the art of being present, of observing thoughts and feelings without judgment, like watching clouds drift across the sky. This simple act can untangle the knots of rumination that tighten the grip of depression. Science backs this up, showing us pictures of the brain literally changing – calming the primal fear centers (the amygdala) and strengthening the regions responsible for thoughtful response (the prefrontal cortex).

Then there's Transcendental Meditation (TM), the quiet repetition of a mantra, guiding the mind into profound rest, dissolving stress, and easing the anxieties that isolation breeds. Or Loving-Kindness meditation, a practice that actively cultivates warmth and compassion – first for oneself, then radiating outwards. Imagine the power of this in a tiny capsule millions of miles from Earth, consciously building bridges of empathy to combat the friction of confinement. These aren't just coping mechanisms; they are ways to actively reshape one's inner world, to find spaciousness even when physically confined.

This knowledge isn't gathering dust on academic shelves. It's being woven into the very fabric of astronaut preparation. Space agencies understand that an astronaut's mind is as critical as their spacecraft's engine. Before they even glimpse the launchpad, candidates are screened and trained, building an emotional toolkit for the voyage ahead. Increasingly, mindfulness and meditation are part of that essential kit, skills they can deploy autonomously when the silence stretches too long.

During the mission, a lifeline remains. Regular, private conversations with psychologists on Earth offer a vital tether. Digital tools, perhaps including guided meditations tailored to their unique situation, provide on-demand support. The HI-SEAS missions themselves become invaluable proving grounds, letting us see in real-time what works, what helps, when support is most needed. And when they return, the journey isn't over; support continues as they readjust, carrying the weight of their extraordinary experience.

This isn't merely about preventing breakdown; it's about enabling astronauts to thrive. It's about ensuring minds stay sharp, teams remain cohesive, and the mission succeeds because the human element – the heart and soul of exploration – remains resilient. The lessons learned in these extreme environments ripple outwards, offering wisdom for anyone facing isolation, whether in Antarctic research stations, submarines, or even during unforeseen quarantines back on Earth.

So, as we reach for the stars, we must also reach inward. The void of space is vast, but the universe within the human mind holds its own mysteries and challenges. By understanding the psychological toll of isolation, illuminated by simulations like HI-SEAS, and by embracing powerful, evidence-based tools like meditation, we equip our explorers not just to endure the journey, but to navigate the cosmos – both outer and inner – with strength, clarity, and unwavering spirit. The frontier is not just "out there"; it is profoundly, unavoidably, within us.

Starlight & Scalpels: Crafting the Future of Health Beyond Earth

For millennia, we've looked up, dreaming of touching the stars. Now, as we stand on the precipice of planting our boots firmly on the Moon and casting our eyes towards the red dust of Mars, that dream feels tantalizingly close. But the cosmos, breathtaking as it is, plays by different rules. It's a realm of stark beauty and silent threats, an unforgiving void where the familiar comforts of Earth's embrace vanish, leaving the human body vulnerable in ways we're only beginning to grasp.

Out there, beyond our planet's magnetic shield, an invisible storm rages – the relentless hail of cosmic radiation. It's not like a bad sunburn; it's energy that slices through flesh and bone, capable of rewriting our very DNA, planting seeds for cancers and diseases decades down the line. Imagine floating through that, knowing the very fabric of space is subtly fraying your internal blueprint. It's a chilling prospect, demanding more than just better shielding or faster ships. It demands we rethink ourselves.

Whispers of Resilience: Can We Edit Our Way to the Stars?

This is where the story takes a turn into the truly audacious. Scientists are looking at life's ultimate survivors – tiny creatures like the near-indestructible tardigrade, the 'water bear' that treats lethal radiation like a minor inconvenience, or bacteria that can piece their shattered genomes back together. What if, they whisper, we could borrow their secrets?

Enter CRISPR, the gene-editing tool that's less a scalpel and more a molecular word processor. The idea, still shimmering on the horizon of possibility and fraught with ethical questions we must navigate carefully, is breathtaking: could we gently tweak our own cellular machinery? Perhaps bolster our natural DNA repair crews, making them faster, more efficient? Or maybe even weave in protective elements inspired by those extremophile masters of survival? It's not about creating superhumans, but about giving our pioneers a fighting chance, crafting internal armor against the cosmic onslaught. It's a profound shift – moving from merely treating radiation sickness to proactively building resilience right into our cells.

The Doctor in the Machine: Our Silent Guardian in the Void

Now, picture the loneliness of deep space. Home is a pale blue dot millions of miles away. A simple radio message takes agonizing minutes to cross the gulf. If disaster strikes – a sudden illness, a critical injury – there's no ER down the hall, no team of specialists on speed dial. Help is literally worlds away.

This is where Artificial Intelligence steps out of science fiction and becomes a lifeline. Imagine an AI, not as a cold calculator, but as a tireless, vigilant partner. It's the digital co-pilot for the human body. Fed data from wearable sensors tracking every heartbeat, breath, and subtle temperature shift, it learns the unique rhythm of each astronaut. It analyzes scans – an ultrasound revealing a hidden anomaly, a microscopic view of a blood sample – with eyes that never tire, cross-referencing findings against vast medical knowledge in milliseconds.

Think of it: a crew member performs a scan, guided by the AI's calm interface. Instantly, the AI flags a concern, offers potential diagnoses ranked by probability, maybe even suggests initial treatment protocols based on the astronaut's specific profile and the ship's medical supplies. It can pre-digest complex data, highlighting the critical points for those delayed consultations with Earth, turning a 40-minute communication lag from a terrifying barrier into a manageable hurdle. Systems like CIMON on the ISS are the first steps – AI companions learning to work alongside humans. The next generation will be true medical guardians, always on watch, embedded within the very habitat.

Weaving the Threads: A New Tapestry of Health for the Frontier

Combine these threads – the potential for enhanced biological resilience and the watchful intelligence of AI – and a new vision of healthcare emerges. It's not just about fixing what breaks; it's about continuous optimization, prediction, and prevention.

The Body's Subtle Language: AI listens constantly to the body's whispers via sensors, catching the earliest signs of trouble – radiation exposure, infection, stress – long before an astronaut even feels unwell. It recommends personalized adjustments: tweak the lighting, alter the exercise regimen, suggest a specific nutrient boost.

Guidance When Worlds Apart: AI becomes the expert in the room, guiding minimally trained crew through complex procedures using smart instruments, turning a daunting medical emergency into a manageable, step-by-step process.

Printing Parts on Demand?: Look further ahead, and you see technologies like 3D bioprinting joining the ecosystem. Imagine printing skin grafts for a burn, or tissue patches for research, manufactured right there in microgravity – a cosmic workshop for the body itself.

This journey into space medicine is forcing us to be bolder, more imaginative than ever before. We're learning to defend the human body against the universe itself, using tools drawn from the code of life and the logic of machines. It's a monumental undertaking, pushing the very definition of medicine. And the incredible part? The solutions we craft for the silent void – the AI diagnostics, the genetic insights, the models for remote, autonomous care – won't stay in space. They promise to echo back to Earth, potentially transforming healthcare for all of us living beneath the stars.

AI as Space's Silent Partner

Our Minds Among the Stars: Machines That Dream of the Cosmos

Imagine whispering instructions across a gulf of millions of miles, only to wait an agonizing half-hour for even a simple nod of understanding. This is the reality of reaching for Mars. Now, imagine guiding a needle into a thread while both are moving faster than speeding bullets. That's the challenge of docking with a space station. For generations, these cosmic hurdles seemed insurmountable, tethering our ambitions to the slow pace of long-distance commands. But something incredible is happening. We're not just sending machines into the void anymore; we're sending extensions of our own ingenuity, machines imbued with a spark of autonomous thought, capable of navigating the unknown and performing delicate cosmic ballets far beyond our real-time reach.

Meet Perseverance, our tireless geologist rolling across the ochre plains of Mars. It's more than a rover; it's a pioneer with digital eyes and a mind of its own. Left to ponder the treacherous beauty of Jezero Crater, 13 light-minutes from its creators, Perseverance can't afford to wait for directions. Using its stereo vision, it drinks in the landscape, painting 3D pictures in its electronic consciousness. It doesn't just see rocks and slopes; it understands hazards – the wheel-trapping sand, the precarious ledge, the mission-ending boulder. Its algorithms, descendants of earlier Martian explorers but far sharper, weigh countless potential paths, choosing the safest, smartest route forward like a seasoned mountaineer picking a line up a complex ascent. This isn't just following a map; it's reading the terrain, thinking on its wheels. And it remembers its journey, constantly comparing snapshots of the ground it just covered, allowing it to know precisely where it is, even when the rusty soil slips beneath its treads – a digital breadcrumb trail ensuring it never truly gets lost. This freedom to think, to decide, allowed it to land with pinpoint accuracy, matching live views to stored maps in a breathtaking, automated plunge, and now lets it roam further and faster than any Martian explorer before, transforming silent landscapes into troves of discovery.

Now, pivot your gaze from the red dust to the star-dusted blackness of Earth orbit. Here, another kind of thinking machine performs a different, equally vital task. SpaceX's Crew Dragon, often carrying precious human cargo, approaches the International Space Station not just with powerful engines, but with a focused, digital intellect. Docking is a dance of infinitesimal precision. Dragon doesn't just blunder towards the station; it uses laser eyes (LiDAR) to feel the distance, thermal sensors to sense the station's presence, and cameras to align itself with microscopic accuracy. Its electronic brain – its guidance, navigation, and control system – acts as a master choreographer. It constantly analyzes its position, velocity, and orientation relative to the speeding station, whispering commands to its thrusters, nudging itself gently, perfectly, along an invisible corridor in space. This isn't brute force; it's finesse born of calculation. And crucially, it possesses a digital instinct for safety. If anything feels wrong – a slight misalignment, an unexpected closing speed – it doesn't hesitate. It automatically backs away, holding steady, ensuring the safety of the crew and the station before proceeding. It's a silent, vigilant co-pilot, handling the intense pressure of arrival with cool, calculated grace.

These two explorers, Perseverance and Crew Dragon, are not mere puppets dancing on interplanetary strings. They are the vanguard of a new era. They embody the solution to the tyranny of distance, bridging the silent void where human commands lag hopelessly behind. They are efficiency multipliers, working tirelessly, making split-second judgments that maximize scientific return or ensure a safe arrival. They are digital guardian angels, spotting hazards or executing critical maneuvers with unwavering precision, reducing the potential for human error in high-stakes environments.

They are also the key that unlocks audacious future dreams: robots collaborating to build bases on the Moon, autonomous craft venturing into the icy depths of the outer solar system, perhaps even bringing samples back from Mars without constant human oversight. Every day Perseverance charts a new course across Mars, every time a Dragon glides seamlessly into its port on the ISS, we witness not just engineering prowess, but the burgeoning intelligence of our creations. Even the little Mars helicopter, Ingenuity, coordinating its flights with Perseverance, offered a glimpse into a future of robotic teamwork across alien worlds.

We are teaching machines to see, to reason, to act, and in doing so, we are extending our own senses and capabilities across the cosmos. These thinking machines are becoming our essential partners, digital envoys carrying our curiosity, resilience, and ambition into the great unknown. The journey has just begun, and as our machines grow smarter, the universe opens wider.

Our New Cosmic Eyes, and the Brilliant Mind We Built to Understand Their Vision

Imagine humanity has just opened a new eye. Not just any eye, but one so powerful it can gaze back to the dawn of time, witnessing the first stars flicker to life and galaxies bloom in the cosmic nursery. That eye is the James Webb

Space Telescope (JWST). It's sending us postcards from the edge of reality – shimmering infrared whispers from ancient suns, the chemical breath of alien worlds, the chaotic beauty of star-forming nebulae. It's breathtaking.

But here's the catch: this cosmic eye is so good, it's showing us everything. It's not sending postcards; it's unleashing a torrential flood of starlight, a digital avalanche measured in petabytes – enough data to fill libraries upon libraries every single year. Hidden within this glorious deluge are secrets we've only dreamed of finding, but spotting them is like trying to find a specific grain of sand in a hurricane by hand. It's simply too much for human eyes and brains alone.

Enter the co-pilot for this incredible journey: Artificial Intelligence. Think of JWST as the most sensitive instrument ever built, capable of hearing the faintest notes of the universe's song. AI, particularly its learning branch (machine learning), is the master conductor we need, capable of picking out the melody from the noise, identifying the individual instruments, and even noticing when a completely unexpected note is played. It's not just helpful; it's become the essential partner in decoding the cosmos JWST is revealing.

This cosmic data isn't just vast; it's exquisitely complex. We're talking about ghostly smudges in deep-field images that could be the most distant galaxies ever seen, or incredibly subtle dips and bumps in light spectra that hint at water vapor on a planet light-years away. AI algorithms, trained on everything we've learned from past explorations (like Hubble's trailblazing views) and sophisticated simulations, are learning to navigate this complexity at lightning speed.

Imagine AI like a tireless cosmic cartographer. We're teaching deep learning networks (a type of AI inspired by our own brain's structure) to recognize the "faces" of galaxies – the elegant spirals, the serene ellipticals, the wonderfully chaotic irregulars – across billions of light-years in JWST's images. They can sort through thousands of possibilities faster than any human team, freeing up astronomers to ponder what these galactic shapes mean for the universe's life story. Similarly, AI is becoming adept at listening to the light from distant atmospheres or glowing nebulae, untangling the overlapping "chemical songs" to tell us precisely what elements are present, even when the signals are faint and jumbled.

But perhaps the most thrilling role for AI is as a "serendipity engine" – our scout for the utterly unexpected. Finding what you're looking for is science; finding what you never imagined is discovery. Remember the Kepler telescope's hunt for exoplanets? While programmed searches found many planets crossing their stars, AI helped sniff out the weird ones – the faint signals, the oddly behaving systems – that didn't fit the standard mold, revealing a richer, stranger planetary zoo than we expected.

Now, unleash that power on JWST's data! We're training AI to learn what "normal" looks like – a typical star, a standard galaxy, a known chemical signature – and then wave a flag whenever it spots something that doesn't fit. A cosmic anomaly. A glitch in the matrix. This could be anything: a type of object we've never conceived of, a bizarre chemical cocktail on an exoplanet challenging all our theories, or maybe even the faint, fleeting signature of the universe's very first, primordial stars (the elusive Population III). AI becomes our tireless watchman, scanning the endless data streams and whispering, "Hey, look at this. This is... different."

Beyond the thrill of pure discovery, AI is also becoming the indispensable stagehand, making the whole cosmic show run smoothly. It helps automatically clean up the raw JWST data – removing cosmic ray 'zaps,' aligning images perfectly, filtering out noise – transforming the raw signals into pristine, science-ready views faster and more consistently than ever before.

Take the hunt for gravitational lenses – nature's own telescopes. Massive objects like galaxy clusters warp spacetime, bending and magnifying the light from even more distant objects behind them. Finding these faint, stretched arcs and multiple images used to require hours of painstaking human searching. Now, AI trained on examples can scan vast JWST images and instantly pinpoint promising candidates, giving us magnified glimpses into the universe's infancy that would otherwise be impossible.

This isn't about replacing human curiosity or intuition. It's about amplification. AI handles the data deluge, the pattern matching, the anomaly hunting at scales we simply can't manage. It frees up human astronomers to do what they do best: ask the big questions, interpret the findings, feel the awe, and dream up the next great cosmic quest.

JWST is writing a new chapter in the story of the universe, scribbled in faint infrared light across vast stretches of time and space. AI is the gifted translator, the tireless collaborator, helping us read that story and understand our place within it. Together, this human-built eye and its artificial brain are embarking on perhaps the greatest exploration ever undertaken – decoding the cosmos, one revelation at a time.

Beyond the Launchpad: Building Our Future, Piece by Weightless Piece

For decades, reaching for the stars meant wrestling with Earth's relentless grip. We crammed our grandest ambitions into the tight confines of rocket noses, like cosmic contortionists folding intricate origami. Every single bolt, every wire, had to survive the fiery punch through the atmosphere and arrive ready-made. It was ingenious, yes, but also profoundly limiting. We were visitors, unpacking carefully prepared suitcases in the void.

But imagine this: what if space wasn't just a destination, but a workshop? What if, instead of shipping fully formed structures, we could send up the raw potential – the cosmic clay – and sculpt our future right there, in the quiet dance of microgravity? This isn't science fiction anymore; it's the revolution quietly humming aboard space stations and being prototyped in labs back on Earth. We're learning not just to visit space, but to build in it.

The Digital Weavers and the Robot Artisans

Think of 3D printers, but unbound from gravity's pull. Companies like Redwire (who absorbed the pioneers at Made In Space) have already proven it's possible. Their Additive Manufacturing Facility on the ISS isn't just churning out spare parts; it's a beachhead. It's the first taste of liberation from Earth-bound supply chains, printing a wrench or a cover on demand, miles above the planet.

Now, infuse that printer with a clever mind – Artificial Intelligence. This AI isn't just following instructions; it's a digital artisan, understanding the strange rules of the orbital workshop. It designs structures born of space, inherently understanding the silent stresses of microgravity, the sun's harsh glare, the whisper of radiation. It watches the printing process with tireless digital eyes, nudging parameters, correcting flaws before they form, ensuring each layer is perfect, even when mission control is a light-second away. Picture Redwire's Archinaut technology: a robotic arm wielding a printer, extruding beams longer than a bus, structures that could never have fit in a rocket, all unfolding in the vacuum like a slow-motion metallic flower. The AI is the conductor, ensuring this delicate ballet of creation proceeds flawlessly, adapting on the fly to the universe's whims.

But printing the pieces is only half the dream. You need the assembly crew. Today, we have marvels like Canadarm2, strong, steady arms guided by human hands from afar. But the future demands more – tireless, autonomous workers. Imagine, as researchers at places like MIT have, small, agile robots – maybe like mechanical spiders or self-assembling cubes – skittering across nascent structures. They wouldn't just hold things; they'd be the carpenters, the electricians, the welders of the high frontier. Picture swarms of them, collaborating like a colony of digital ants, guided by an AI foreman, raising trusses, connecting modules, laying down power lines. They wouldn't need oxygen or sleep, just instructions and energy, tackling complex builds with intricate, coordinated movements.

From Blueprints to Star-Forged Reality

When these two streams converge – AI-driven printing and autonomous robotic assembly – the possibilities truly explode. We're not just talking about bigger space stations anymore.

Think kilometer-wide solar power arrays, unfurling like gossamer wings in orbit, beaming clean energy down to Earth. Imagine telescopes with mirrors so vast they dwarf

Hubble and Webb, capable of peering into the universe's cradle. Launching these behemoths is impossible. Building them in orbit, perhaps even using materials mined from asteroids (talk about cosmic recycling!), becomes achievable. Robots print the struts, other robots connect them, guided by AI overseeing the grand design. NASA's OSAM-1 mission, with its SPIDER robot designed to assemble an antenna in space, is an early step on this path, a practical demonstration of robotic hands doing intricate work far from home.

Consider our future homes among the stars, like the planned Lunar Gateway. Instead of launching entire, heavy modules, maybe we launch the 'seed' and let robotic systems print and attach new rooms, patch up micrometeoroid dings with custom-printed shields, or swap out aging external gear. Fewer dangerous spacewalks for astronauts, more robust, adaptable habitats growing organically in space.

And what about the journey further out? Building massive fuel depots or interplanetary ferry ships piece by piece in Earth orbit or at stable Lagrange points could become the norm. These wouldn't be cramped capsules, but potentially spacious transfer habitats assembled by our tireless robotic crews, making the long voyages to Mars and beyond more feasible, more sustainable.

The Dawn of the Orbital Construction Site

We stand at the cusp of a profound shift. By taking the construction site off-world, we unshackle ourselves from the tyranny of the rocket equation. AI and robotics are becoming the skilled hands and watchful eyes that will allow us to build bigger, bolder, and more beautiful structures in space than ever before. From Redwire's tangible prints on the ISS to the visionary concepts dancing in university labs, the tools are being forged.

This isn't just about engineering; it's about transforming the vacuum from a void to be traversed into a canvas upon which humanity can build its next chapter. It's about safety, sustainability, and unlocking a future where space is not just explored, but truly inhabited and utilized. The silent, vast expanse is waiting. The cosmic construction crew is gearing up.

Our Digital Genie: Taming the Power at AI's Sharp Edge

Artificial intelligence isn't just knocking on the door anymore; it's practically moved in, rearranging the furniture of our lives and industries. It learns faster than we can teach, sees patterns we miss, and makes choices in the blink of an eye. It's dazzling, powerful, and maybe a little terrifying. Because within that incredible power lies a sharp, tricky boundary – what we might call AI's "ethical edge." Walking this edge means understanding the shadows that trail this brilliant light.

Echoes in the Code: When AI Learns Our Flaws

Think of AI decision-making like looking into a complex digital mirror. What happens if the mirror itself is warped? Ethics journals are buzzing with warnings about this very problem. AI learns from the data we feed it, and if that data carries the ghosts of our own societal biases, the AI doesn't just reflect them – it can amplify them into algorithmic prejudice (as scholars like Floridi & Taddeo have explored). It's not abstract theory. Imagine software meant to predict crime hotspots inadvertently painting targets on minority neighborhoods simply because it learned from skewed historical arrest records (a chilling reality documented by the AI Now Institute). Or consider the now-infamous COMPAS system, studied by ProPublica, which tragically acted like a biased judge, wrongly flagging Black individuals as high-risk reoffenders far more often than white individuals. These aren't code errors; they're echoes of human failings embedded in logic.

Then there's the "digital whodunit." When a self-driving car swerves tragically, or a medical AI misses a crucial diagnosis, who carries the weight? The coder who wrote the initial lines? The company that supplied the data? The human who trusted the output? Or the algorithm itself, humming away silently? Our old ideas about blame and responsibility feel suddenly inadequate, like trying to catch smoke. Researchers like Matthias and Sparrow call this the "responsibility gap" – a troubling space where accountability dissolves in the complex dance between human intention and machine action. Add to this the unnerving possibility of AI pursuing its programmed goals in ways we never intended, creating unforeseen harm (Bostrom's "alignment problem") – it's like giving a powerful tool instructions without fully grasping how it might interpret them.

Cracking Open the Crystal Ball: Why We Need to See Inside

If we're going to trust these powerful new minds, we need more than just results; we need understanding. Many advanced AIs operate like impenetrable "black boxes." Their intricate calculations are hidden, their reasoning obscure even to the people who built them. This opacity isn't just inconvenient; it's a barrier to trust, a finding echoed in studies from AI safety labs like OpenAI and DeepMind. How can we fix errors we can't trace? How can we be sure an AI shares our values if we can't glimpse its "thought process"? Research consistently shows that people simply don't trust what they can't understand (a point driven home by Doshi-Velez, Kim, Riber… and Lapedriza).

This is where the quest for Explainable AI (XAI) becom… not just an academic pursuit, but a vital necessity. Ima… a doctor relying on an AI diagnostic tool. They need to know why the AI suspects cancer, not just receive a digita… thumbs-up or down (as emphasized by work like the DARPA XAI program). Transparency is the key to unlocking trust, enabling audits for hidden biases, and spotting vulnerabilities before they cause harm. Without it, we're flying blind.

Cosmic Crossroads: Ethics Beyond Earth's Atmosphere

Now, launch these ethical quandaries into the vast emptiness of space. AI is becoming the silent co-pilot on our cosmic journeys, guiding rovers like Perseverance across Martian plains with superhuman speed (using systems like AutoNav), sifting through mountains of telescope data, and managing precious resources light-years from home.

But this interstellar reliance forces us to ask profound questions. What rules govern an AI deciding which scientific experiment gets priority when oxygen runs low on a future Mars base? Could an AI trained on Earthly life accidentally overlook a truly alien biosignature because it doesn't fit the mold? (These aren't sci-fi hypotheticals; NASA's AI initiatives wrestle with them now).

Transparency here is literally mission-critical. Ground control needs to understand why a probe suddenly changed course during a communications blackout that stretched across millions of miles. Verifying an AI's decision after the fact is crucial for safety and learning, especially when dealing with high-stakes maneuvers like autonomous docking (a focus for the European Space Agency). We need cosmic governance – rules for data purity, bias checks tailored for the unknown, clear validation steps, and accountability protocols designed for situations where help is not an option. A mistake isn't just costly; it could be catastrophic, measured in lost missions or, one day, human lives adrift in the void.

So, navigating AI's "ethical edge" is about embracing the incredible potential while staying perpetually vigilant about the risks. It's about demanding transparency, questioning the data, and designing systems that reflect our best intentions, not our buried biases. From the algorithms shaping our daily lives to the digital minds exploring the stars, building trustworthy AI requires constant dialogue, rigorous research, and a shared commitment to guiding this powerful creation wisely. It's less a finish line, more a horizon we must continually strive towards.

Clearing the Orbital Highways

Our Crowded Sky: More Than Just Starlight

Lean back tonight, cast your eyes skyward. That vast, velvet expanse whispers serenity, emptiness. But hold that thought. Just beyond our breathable air, we've drawn a messy halo around our planet – an invisible, high-velocity junkyard buzzing with the ghosts of missions past. The official count from cautious space agencies like ESA? A mind-boggling 128 million pieces of stuff bigger than a millimeter.

Imagine that. It's not just stars up there. It's a ghostly census including around 36,500 trackable chunks – dead satellites like orbiting tombstones, discarded rocket bodies like giant metal husks, and sharp fragments from past smash-ups, all diligently watched by blinking eyes on Earth. But the real terror lies in the shadows: a million more pieces are lurking between the size of a marble and a softball, and then the staggering 128 million down to the scale of a grain of sand.

"So what?" you might ask. "A fleck of paint?" Here's the cosmic kicker: speed. These aren't drifting aimlessly. They're screaming around Earth at nearly 28,000 kilometers per hour (that's almost 8 kilometers every single second!). At that velocity, physics gets brutal. That tiny fleck transforms into a hypervelocity projectile. A simple 1-centimeter aluminum ball hits with the energy of a bowling ball hurtling down a highway at 100 km/h. Scale up to something 10 cm across, and it's not a collision; it's an obliteration. Suddenly, 128 million isn't just a number; it's 128 million potential bullets in a game of orbital Russian roulette we started.

The Kessler Cascade: When Space Runs Out of Space

Now, picture this orbital junkyard getting thicker. That's the chilling prophecy known as the Kessler Syndrome, envisioned by NASA scientist Donald Kessler way back in 1978. He warned of a tipping point, a moment where there's so much junk that collisions inevitably happen. But here's the twist: each collision doesn't remove debris; it creates vastly more, smaller, harder-to-track pieces. It's a runaway chain reaction, a self-perpetuating demolition derby turning useful orbits into impassable minefields.

Think cosmic billiards, but every time two balls hit, they shatter into a thousand more, each just as fast, dramatically increasing the odds of hitting everything else. Scientists run complex simulations – digital crystal balls like NASA's LEGEND and ESA's MASTER – trying to map this potential future. While the "when" is fuzzy, the "if we do nothing" is alarmingly clear: vital orbital highways could become unusable, potentially for generations.

This isn't just theory. We've had terrifying previews. In 2009, an old Russian satellite (Cosmos-2251) slammed into an active Iridium communications satellite. Boom. Thousands of new, dangerous pieces instantly littered orbit. And in 2007, a deliberate Chinese anti-satellite test shattered a weather satellite, single-handedly creating a massive, long-lasting debris cloud we still navigate around. These weren't hypotheticals; they were cosmic fender-benders with catastrophic potential, stark previews of the Kessler nightmare.

The Daily Dodge: Living with the Junk

So, how do we live in this increasingly cluttered cosmic backyard? Very, very carefully. The knowledge of those 128 million threats and the looming Kessler Syndrome isn't just academic chatter; it dictates survival for multi-billion dollar satellites and, more importantly, the astronauts living in space.

Cosmic Traffic Control: Imagine tireless watchdogs – radar dishes and telescopes scattered across the globe, primarily led by the US Space Force – constantly scanning the skies. They track those 36,500+ larger objects, predicting near-misses (conjunctions). When a potential collision path looks too close for comfort, alarms ring.

The ISS Sidestep: Our home in orbit, the International Space Station, lives this reality daily. NASA and its partners are constantly playing orbital dodgeball. If a tracked piece of junk threatens to enter the station's imaginary "safety bubble," mission control fires the station's thrusters (or those of a docked ship) in a carefully choreographed Collision Avoidance Maneuver (CAM) – a slight nudge up, down, or sideways to get out of harm's way. They've had to do this dozens of times. Each maneuver is a stark reminder of the hazards floating just outside the window.

Automated Agility: For massive satellite constellations like Starlink or OneWeb, with thousands of individual satellites, manual dodging is impossible. They rely on sophisticated automated systems – think AI-powered space traffic controllers – constantly crunching tracking data, predicting threats (from debris and other satellites), and commanding tiny, autonomous course corrections. It's an intricate, automated orbital ballet performed non-stop.

Smarter Design, Safer Skies: Knowing the risks shapes the future. New missions might aim for less crowded orbital paths. Spacecraft are built with some shielding (though it's like wearing a t-shirt against a bullet for larger debris). Crucially, there's a growing commitment to responsibility: designing satellites to naturally fall and burn up within 25 years, or saving enough fuel to actively steer them into the atmosphere at mission's end.

That number – 128 million – isn't just a statistic. It's the context for every space mission launch, every satellite maneuver, every astronaut's safety plan. It represents the pressing need to not only track the mess we've made but to actively start cleaning it up and, more importantly, stop adding to it so carelessly. The challenge isn't just technical; it's about stewardship. Will we tend to our cosmic backyard, or will we allow our vital orbital highways to become impassable ruins, locking us onto our planet beneath a ceiling of our own making?

Catching the Cosmic Draught: How Ghostly Kites and Smart Design are Cleaning Up Our Orbital Backyard

Imagine looking up, not just at the stars, but at the invisible highways buzzing with the satellites that power our modern world. Now, imagine those highways getting dangerously cluttered with silent, speeding space junk. It's a real problem, born from decades of exploration. But fear not! Brilliant minds, like the teams at Surrey Satellite Technology Ltd (SSTL) and the Surrey Space Centre (SSC), are crafting ingenious solutions – not with cosmic tow trucks, but with something far more elegant: giant, gossamer sails.

The Whisper-Thin Parachutes of LEO

Think of these de-orbiting sails not as engines, but as vast, ultra-light parachutes designed for the near-vacuum of Low Earth Orbit (LEO). Up there, even the thinnest wisp of atmosphere creates drag, a tiny friction that wants to pull things down. For a sleek satellite, this pull is negligible, meaning it could haunt our skies for decades, even centuries, after its job is done.

But unfurl a massive sail – sometimes using clever inflatable booms – and suddenly that satellite is like a kite catching the faintest cosmic draught. Its surface area balloons, dramatically increasing the atmospheric drag. Physics then does the heavy lifting: more drag means the orbit shrinks, faster and faster, until the satellite takes a final, fiery plunge back through the denser atmosphere, burning up harmlessly into dust. It's a graceful, propellant-free exit strategy. Higher up, where the air is truly gone, these same sails can even catch the incredibly subtle push of sunlight itself – photons acting like a gentle, persistent breeze to nudge satellites towards their final destination, whether that's re-entry or a safe "graveyard" orbit far from the busy lanes.

Surrey's Sky-High Experiments: Proof in the Plunge

This isn't just theory. Surrey and their collaborators have sent these ideas soaring:

InflateSail's Dramatic Dive: Picture this: a tiny 3.2 kg satellite, launched in 2017 as part of a bigger mission. Left alone, it would have circled Earth for years. But then, it deployed its 10 square-meter sail. The result? Instead of lingering, it plunged back to Earth in a mere 72 days. Tracking data didn't lie; this ghostly kite worked spectacularly well, providing undeniable proof that you can dramatically accelerate a satellite's return trip.

RemoveDEBRIS's Bigger Test: This ambitious 2018 mission was like a Swiss Army knife for tackling space junk, testing harpoons and nets. But crucially, after its main tasks, the ~100 kg platform deployed its own sail. While its journey home took longer (it started higher and had other jobs to do), mission control watched as its orbital decay rate measurably sped up the moment the sail unfurled. It confirmed the principle worked not just for tiny cubesats, but for heftier platforms too.

These sails are appealing because they're beautifully simple once open – no fuel needed, just physics. They're light, pack down small, and offer a smart way to meet global guidelines urging us to clear out old satellites within 25 years. Especially for constellations of smaller satellites, they're a game-changer.

Bringing the Big Birds Home: Starship and the Reusable Revolution

For decades, one of the biggest sources of dangerous space junk wasn't dead satellites, but the massive upper stages of rockets – the delivery trucks left abandoned after dropping off their precious cargo. Enter the era of reusable rockets, with SpaceX's Starship leading the charge.

Starship isn't just a rocket; it's designed to be more like an airliner for space. Both the giant booster and the Starship upper stage (the part that actually goes to orbit) are built to fly back, land, and launch again. This is radically different. Instead of discarding a multi-ton stage in orbit after deploying satellites, the plan is for Starship to fire its engines one last time, point itself home, and glide back to a landing pad.

By its very design, Starship aims to prevent these enormous potential debris pieces from ever being created. It tackles the "abandoned rocket body" problem head-on. Imagine launching hundreds of tons to orbit, and then bringing the launch vehicle back. If this becomes the norm (with others like Blue Origin's New Glenn following suit), it could fundamentally slash the rate at which we add large, dangerous objects to the orbital environment. Starship doesn't solve the satellite end-of-life issue itself, but it stops the delivery truck from becoming part of the junkyard.

Building Tidiness In: The Art of the Responsible Satellite

Beyond sails and returnable rockets, the cleanest path forward involves designing satellites with their entire lifecycle – including their demise – in mind right from the drawing board.

Pulling the Plug Safely (Passivation): Think of it like making a derelict building safe. Before abandoning a satellite, operators vent leftover fuel, discharge batteries, and release any pent-up pressure. Why? Because a "dead" satellite full of stored energy is a ticking time bomb. A random collision or component failure years later could cause it to explode, creating thousands of tiny, hard-to-track shrapnel pieces – the worst kind of debris. It's a critical housekeeping step.

Designing for Disappearance (D4D): This is about making satellites that want to burn up completely on re-entry. Engineers are cleverly choosing materials and structural designs that break apart and vaporize more easily in the intense heat. Using materials that melt at lower temperatures or designing structures to peel apart ensures that little, if anything, survives the fiery descent to potentially hit the ground. It's like designing biodegradable packaging for space hardware.

Packing Your Own Exit Plan: Increasingly, satellites are launching with their own built-in end-of-life plan. This might be a small reserve of fuel for a final de-orbit burn, an efficient electric thruster to gently spiral down over months, or, yes, a neatly folded de-orbit sail ready to deploy when the time comes. Major constellation operators are already building this in, programming their satellites to autonomously dodge collisions and actively de-orbit at mission's end. It's about taking responsibility from cradle to grave (or rather, fiery re-entry).

Weaving a Cleaner Future in Orbit

There's no single magic bullet for the space debris challenge. It demands a tapestry woven from multiple threads: the passive elegance of de-orbit sails pioneered by groups like Surrey, the paradigm shift of reusable launchers like Starship refusing to leave trash behind, and the ingrained responsibility of designing satellites that know how to say goodbye gracefully.

The evidence from missions like InflateSail and the bold vision of Starship show these aren't just futuristic dreams. They are practical, vital steps we can take. By embracing this toolkit of clever ideas and responsible habits, we can strive to keep the pathways to the stars clear, ensuring that space remains a domain of discovery, connection, and opportunity for all the generations yet to look up in wonder.

Our Littered Heavens: From Cosmic Fishing Nets to Laser Whispers

Imagine looking up, not just at stars, but at a glittering, dangerous halo of our own making – a scrapyard orbiting Earth at mind-boggling speeds. Space, the final frontier, is also becoming our highest junkyard. But pioneers are looking up, not in despair, but with blueprints and bold ideas. This isn't just about engineering; it's about taking responsibility for the cosmic neighborhood we're starting to inhabit.

The Great Orbital Snag: Casting a Net Among the Stars

Think of it as the ultimate fishing trip, but instead of water, you've got the vacuum of space, and instead of fish, you're hunting rogue chunks of metal. That's essentially what the folks at the University of Surrey's Space Centre and their partners like Airbus dared to do with RemoveDEBRIS. Launched from the International Space Station back in 2018 – our orbiting outpost literally tossing out the gear – this wasn't about hauling in mountains of trash just yet. It was more like learning how to fish in zero gravity.

The star moment came in September that year. RemoveDEBRIS released its own practice target – a little satellite pretending to be a piece of dangerous junk. Then, from about 7 meters away (imagine parking a car with inches to spare, but in space!), the main satellite took aim and fired its net. Picture a high-tech spiderweb, about 5 meters across, shooting through the void. And... swoosh! It bagged the target, just like they'd hoped. It sounds simple, maybe, but proving you can actually snare a tumbling, uncooperative object with a net in the weird physics of orbit? That was huge. It was a cosmic "Eureka!" moment, showing that this seemingly simple idea – a net – could actually be a tool in our future space-janitor kit. They didn't just catch junk; they caught invaluable data about how things move, how nets unfurl, and how darn tricky it is to aim right when everything's floating.

The Light Touch: Zapping Junk with Laser Precision

Now, shift gears from physical snags to something straight out of science fiction: using lasers to gently nudge space debris out of the way. Forget explosive blasts; think more like a cosmic nudge, a whisper of light with surprising force. This is laser ablation.

Imagine a powerful, focused beam of light hitting a piece of space junk. That intense energy instantly vaporizes a tiny puff of the debris's surface, creating a miniature jet plume. Like the gentlest of rocket engines, this puff gives the debris a tiny push in the opposite direction. Do this repeatedly, precisely, and you can subtly alter its path. The goal? To gently guide it towards a fiery end in Earth's atmosphere or steer it into a safe "graveyard orbit" where it won't bother anyone.

Researchers, particularly in Japan at places like the RIKEN institute, are deep in exploring this. They're figuring out the perfect laser 'recipe' – the right power, the right light frequency – to effectively vaporize bits of aluminum or satellite insulation. They test this in vacuum chambers on Earth, zapping materials to measure the push they get. Of course, building a space laser powerful enough, aiming it accurately over vast distances, and powering it is a monumental task – like trying to hit a speeding bullet with a laser pointer from miles away, only harder. But the dream of a contactless cleanup method, one that doesn't need to physically grab the debris, keeps scientists chasing this beam of light.

The Big Cleanup: Why Tidying Orbit is So Darn Hard (and Expensive)

Okay, so we've tested nets and we're dreaming of lasers. Why aren't we already scrubbing the skies clean? Because the scale of the problem is staggering. We're talking about tens of thousands of objects bigger than a softball, maybe a million marble-sized pieces, and potentially over 130 million tiny flecks – all whipping around Earth faster than speeding bullets (over 28,000 km/h!). Hitting even a small paint chip at those speeds can cripple a multi-million dollar satellite we rely on for GPS, weather forecasts, or cat videos.

Actually cleaning this up faces some sky-high hurdles:

The Price Tag: Sending anything to space is wildly expensive. Sending up dedicated cleanup missions? Even more so. Convincing someone to pay for orbital garbage collection when the "benefit" is preventing a potential future disaster is a tough sell. Trailblazing companies like Astroscale (practicing magnetic docking with its ELSA-d) and ClearSpace (contracted by ESA to grab an old rocket part) are trying to build the business case, proving it can be done and might eventually pay off.

Tricky Tech: That net worked once, on a known target. But space junk comes in all shapes and sizes, often tumbling unpredictably. We need a whole toolkit – nets, harpoons (also tested by RemoveDEBRIS!), robotic arms – that are reliable, maybe even autonomous, and can handle the brutal cold, heat, and radiation of space without failing.

Who's First? With so much junk, where do you even start? Experts tend to focus on the biggest, heaviest pieces (like old rocket stages) in the most crowded orbital highways, as these pose the biggest risk of creating even more debris if they collide. Getting everyone to agree on the "most wanted" list is another challenge.

Cosmic Law & Order: Who owns that defunct satellite? Who's liable if a cleanup mission accidentally bumps into something valuable, or creates more debris? We need international agreements, clear rules of the road for space cleanup, before things get messy legally and politically.

So, the path forward isn't just one magic bullet (or net, or laser). It's likely a mix-and-match approach, using different tools for different trash. More importantly, it's also about stopping the problem at the source – designing satellites that can safely de-orbit themselves when they're done, and managing space traffic better so things don't bump in the first place.

Turning our orbital junkyard back into a sustainable frontier is a long, complex haul. But from that first successful net toss to the ongoing laser experiments, humanity is slowly, painstakingly learning how to become responsible custodians of the space above our heads.

Earth's Guardian Satellites

Earth's Fever Dream: Our Watchers in the Void

Trying to grasp climate change from ground level alone is like listening to a symphony through a single crack in the wall. You hear something, but the grandeur, the intricate harmonies, the looming crescendos? Lost. We need the full view, the cosmic perspective. That's where our orbital companions come in – not just machines, but extensions of our own senses, soaring above, keeping an unwavering vigil.

Imagine them, gliding silently through the black, their gaze fixed on the swirling blue and white marble below. They drift over sun-scorched deserts where no human stands guard, sweep across the vast, lonely oceans, and witness the silent crumble of ancient ice sheets. They aren't just seeing; they are measuring, taking the planet's pulse, charting its fever, translating the subtle sighs and dramatic shifts of our world into a language we can understand. They are the unbiased witnesses, providing the irrefutable evidence we need to navigate the profound changes underway.

Subtopic 1: The Carbon Confidante - Listening to Earth's Breath

Carbon dioxide – CO2. It's the invisible blanket we keep thickening around our world, trapping heat, altering everything. But where does it really come from? Where does our weary planet manage to stash some of it away? To untangle this, we sent up detectives, chief among them NASA's Orbiting Carbon Observatory-2 (OCO-2), our "Carbon Confidante" launched in 2014.

Think of OCO-2 not just as a satellite, but as having an incredibly discerning eye for light. It watches sunlight bounce off Earth and analyzes the unique way CO2 molecules "color" that reflected light, like reading a hidden message. This isn't just about how much CO2 there is; it's about painting a global map of its ebb and flow. OCO-2 can pinpoint the planet's heavy exhales – the plumes from bustling cities, the ghosts of burning forests – but also its desperate inhales, the deep breaths taken by thriving jungles and hungry ocean waters.

The stories OCO-2 tells are captivating. It revealed how massive El Niño events can make tropical lands gasp, temporarily weakening their ability to soak up our carbon emissions and causing atmospheric levels to spike faster. It confirms, alongside steadfast ground stations like Mauna Loa, the stark reality: our atmosphere is saturated with CO2 at levels unheard of in nearly a million years, pushing past 420 parts per million – a number that should echo like a warning bell. Its successor, OCO-3, perched on the International Space Station, continues this vital surveillance, even zooming in on specific cities, holding up a mirror to our collective footprint.

Real-world Echo: Imagine California, striving to curb its emissions. Insights gleaned from the sky, akin to OCO-2's vigilance, help officials see if their efforts are truly working, region by region, sector by sector. It's space-borne accountability, ensuring promises made on the ground are reflected in the air we breathe.

Subtopic 2: Charting the Creeping Flood - The Jason Legacy

Feel the warmth? It's not just the air. Our oceans are absorbing the vast majority of the extra heat, and warmer water physically expands. Add to that the meltwater pouring from glaciers and ice sheets – the tears of a warming world – and you get the slow, inevitable rise of the sea. Measuring this global phenomenon demands incredible precision, a task entrusted to our orbital surveyors, the Jason satellite family – a testament to international collaboration (USA, France, Europe).

Jason-3, launched in 2016, carried the torch from predecessors dating back to 1992, meticulously charting the ocean's surface. How? Imagine it bouncing a radar ping off the sea and timing the echo's return with astounding accuracy. Knowing precisely where the satellite is allows scientists to calculate the sea's height down to the millimeter. It's like having a supremely accurate measuring stick reaching down from space.

The decades-long record kept by Jason and its kin (now including Sentinel-6 Michael Freilich) is unequivocal: the sea isn't just rising; it's accelerating. The global average creep is now around 3.4 millimeters per year – adding up to over 10 centimeters (4 inches) since the early '90s. But this isn't a uniform bathtub filling up. Jason's data reveals a complex tapestry – some coastlines are tragically sinking much faster than others, victims of currents, sinking land, and the sheer weight of this global change.

Real-world Front Lines: Picture the people of Tuvalu or the Marshall Islands. For them, Jason's data isn't academic; it's a forecast of their future, dictating where to build sea walls, how to plan communities, and sometimes, confronting the heartbreaking possibility of relocation. This isn't just data; it's a lifeline, informing survival strategies on the front lines of climate change and underpinning pleas for international support.

From Orbit to Action: Weaving Data into Decisions

These whispers from orbit, these meticulous measurements, are the bedrock upon which we must build our response:

Sharpening Our Glimpse into Tomorrow: Climate models, our complex digital simulations of Earth's future, feast on this satellite data. OCO-2's carbon maps and Jason's sea level trends make these models less like hazy crystal balls and more like reliable forecasts, guiding our preparations.

Holding Nations to Their Word: Global promises, like the Paris Agreement, rely on honesty. Satellites offer a powerful tool for verifying reported emissions, shining a light on major sources and ensuring accountability on a global scale.

Drawing Lines Against the Flood: Coastal communities worldwide use Jason's hard numbers. Engineers designing defenses, planners zoning new developments, emergency teams preparing for storm surges (made worse by higher seas) – they all rely on this orbital perspective to make life-saving decisions.

Taking Our Temperature: As we implement policies, how do we know if they're working? Continuous monitoring from space provides the feedback loop, showing us if our efforts are truly bending the curve.

These satellites, our silent partners in the sky, do more than just observe. They transform abstract threats into tangible realities. They provide the objective, global truth needed to understand the scale of the challenge, to assign responsibility, to anticipate the future, and crucially, to guide our collective hand as we strive to heal our planet and adapt to the changes already reshaping our world. They are our eyes, our measures, our undeniable witnesses.

Seeing the Unseen: Earth's Portrait Painted with Light and Whispers from Orbit

Imagine trying to understand the intricate details of a vast, living sculpture – our Earth – using only broad strokes. For years, that's how we often saw our planet. But now, we have tools that are less like blunt brushes and more like the finest mapping pens and tireless, watchful eyes. Meet lidar and CubeSats, two revolutionary ways we're getting to know our world with breathtaking intimacy.

Lidar: The Laser Whisperer Revealing Earth's Secrets

Think of lidar not just as "light detection and ranging," but as a kind of hyper-precise echo. It sends out whispers of laser light and listens intently for their return, timing the journey down to the nanosecond. From these fleeting echoes, it builds stunningly detailed 3D portraits – "point clouds" – that reveal the true shape of the land, the texture of a forest canopy, even the hidden contours beneath coastal waters (Result 1.1, 4.2). Born in the 60s, this tech truly came alive when paired with GPS and motion sensors, transforming it into a pinpoint-accurate geographical artist (Result 1.1).

NOAA, the guardian of our oceans and atmosphere, wields lidar like a diagnostic tool. On the coasts, it meticulously maps shorelines, revealing vulnerabilities to rising seas and storm surges, essentially letting us read the coastline's diary of change and glimpse its future (Result 1.2, 4.1, 4.3). High above, lidar probes the air we breathe, profiling invisible dancers like aerosols, water vapor, and ozone, and even charting the winds' hidden currents – vital clues for understanding weather and climate shifts (Result 1.3). And this artist is constantly refining its technique, becoming smaller, more affordable, and even learning to see in multiple "colors" of light (multi-spectral lidar) to not just map shape, but identify what it's seeing (Result 1.2, 6.1).

CubeSats: A Democratization of the Heavens, An Army of Tiny Watchtowers

While lidar dives deep into detail, CubeSats offer a complementary superpower: persistence. Picture a swarm of smart, miniature sentinels, no bigger than a shoebox, circling the globe (Result 2.3, 5.2). Built with accessible, off-the-shelf parts, they've slashed the cost of entry to space, unleashing a wave of innovation (Result 5.1, 5.2). Suddenly, universities and nimble startups – not just space-faring giants – can deploy their own eyes in the sky. This isn't just a trend; it's a booming revolution in how we monitor our planet, a market blossoming towards billions (Result 2.4).

And these little satellites are doing big things for our climate and environment:

OroraTech's Fireflies: A German startup's network literally watches for the heat of nascent wildfires across millions of hectares, offering precious early warnings that ground-based methods might miss (Result 2.2).

vorteX-io's River Readers: From France, this team uses satellite insight (likely including CubeSats) to track the pulse of rivers – their height and flow – giving us foresight into potential floods and droughts (Result 2.2).
Planet Labs' Daily Diary: Imagine photographing nearly every inch of Earth's landmass, every single day. That's Planet Labs' reality, a constant stream of images revealing deforestation, agricultural shifts, and countless other planetary changes, now increasingly analyzed by AI (Result 5.1, 5.2).

Students Reaching for the Arctic: Even university teams are launching sophisticated CubeSats, like Denmark's DISCO-2, planning to use onboard AI and cameras to study the fragile Arctic climate, proving space is truly opening up (Result 5.3).

The Symphony of Sight: When Precision Tools Work Together

The real magic happens when these different perspectives combine. Imagine:

Listening to Forests: Drone-borne lidar maps the precise height and density of trees (Result 3.2, 6.3), while multispectral eyes (from CubeSats or drones) identify species and gauge their health with remarkable accuracy (Result 6.2, 6.3). It's like conducting a detailed census and health check-up for entire forests.

Defending Coastlines: Lidar's detailed elevation maps feed directly into models showing how rising seas or floods might impact communities, turning abstract threats into tangible guides for action (Result 4.3).
Whispering to Crops: Lidar can measure plant growth down to the centimeter, helping farmers use resources wisely and predict harvests (Result 3.2). This ground-truth detail perfectly complements the broader view of crop health provided by CubeSats.

Responding to Crises: OroraTech's CubeSats spot the fire (Result 2.2), lidar maps the damage afterward (Result 1.1), and hydrological networks track the floodwaters (Result 2.2) – a coordinated dance of data for better disaster response.

The Horizon Ahead: Smarter Eyes, Deeper Understanding

The next act? Fitting powerful lidar sensors onto nimble CubeSats, merging deep detail with global reach (Result 3.1, 3.3, 6.4). Add layers of AI to instantly interpret the flood of incoming data (Result 3.1, 5.1), and we're entering an era of unprecedented planetary awareness.

In essence, lidar and CubeSats are more than just technologies; they are extensions of our senses, allowing us to perceive our world with newfound clarity and urgency. They are the precision instruments we desperately need to understand the complex changes underway and navigate towards a more sustainable, resilient future for the living sculpture we call home.

Quantum Leap into Space

Our Cosmic Dreams Need Quantum Minds: Rethinking the Stars

The universe whispers secrets in a language we're only just beginning to learn: quantum mechanics. It's a realm of mind-bending rules governing the impossibly small, yet it underpins the entire cosmic tapestry, from the fiery birth of stars to the silent orbits of satellites. Our trusty classical computers, brilliant as they are, often stutter when faced with this deep, weird reality. They're like trying to conduct a symphony with just a drum. But what if we could build minds that think in the universe's native tongue? Enter quantum computing – not just faster machines, but entirely new kinds of thinkers, "quantum brains" poised to help us truly grasp the heavens.

1. Untangling the Celestial Traffic Jam: Quantum Choreography in Orbit

Picture this: thousands of satellites whizzing around Earth, a complex ballet requiring constant vigilance. Plotting their paths to avoid collisions, ensuring they blanket the globe perfectly, all while sipping fuel – it's a scheduling nightmare that makes rush hour look simple. For today's computers, it's like trying to untangle a million knotted threads simultaneously; the possibilities explode beyond comprehension. They get bogged down checking option after option, one painstaking step at a time.

Quantum computers, however, are the ultimate multitaskers. Thanks to the weird magic of superposition, they can explore countless potential routes at the same time, like seeing every possible path through a maze instantly. Using entanglement, where quantum bits are mysteriously linked, they can weigh complex trade-offs in ways that feel almost intuitive. Imagine algorithms like QAOA or VQE acting as cosmic choreographers, finding elegant, efficient paths through the orbital dance that classical methods might never stumble upon.

The Heart of the Matter: We're not quite there yet with giant, flawless quantum machines. But even today's "toddler" quantum devices are showing glimpses of this power on smaller scales, proving the concept. Think about the sheer scale: guiding even a modest satellite fleet involves more choices than atoms in the known universe! We need a quantum leap.

Imagine the Scene: A sudden debris cloud threatens a vital constellation like Starlink. A quantum optimizer could instantly reroute dozens of satellites, weaving them through danger like cosmic F1 drivers. Or picture plotting a grand tour of the outer planets, using gravity assists like celestial slingshots – a quantum brain could find ingenious, fuel-saving trajectories we haven't even dreamed of, charting truly new paths into the deep dark.

2. Asking the Universe Its Deepest Secrets: Quantum Whispers from the Edge

Where do the laws of the universe seem to fray? At the crushing heart of a black hole, or in the searing chaos of the Big Bang. Here, gravity's immense power dances with the spooky rules of quantum mechanics. Trying to simulate these ultimate extremes on a normal computer is like trying to describe a hurricane using only haikus – you lose the essential chaos and complexity. You can't just approximate quantum weirdness when it is the story.

But quantum computers are quantum systems. They're built from the same stuff they're trying to simulate. Imagine using qubits – the heartbeats of a quantum computer – to mimic the behavior of particles shimmering at the edge of a black hole's event horizon. Could we finally see Hawking radiation, that ghostly particle glow predicted decades ago? Could we unravel the information paradox – does anything truly escape a black hole's grip? Quantum simulations could let us "listen" to the quantum whispers near singularities or replay the universe's first fractions of a second, watching how the seeds of galaxies emerged from the quantum foam.

The Heart of the Matter: Simulating even a sliver of quantum reality takes immense quantum power. Today's efforts are like sketching blueprints, testing methods on "toy" universes. The dream – simulating cosmic events in full quantum glory – might require millions of perfectly stable quantum bits, a distant peak we're still climbing towards.

Imagine the Scene: Physicists huddle around a quantum computer, watching a digital twin of a black hole slowly evaporate, testing fundamental theories of quantum gravity in a way never before possible. Or perhaps they're simulating the primordial soup just after the Big Bang, comparing the quantum computer's output directly to the faint, ancient light mapped by our telescopes, decoding the universe's baby pictures.

3. Drinking from the Cosmic Firehose: Quantum Eyes for Infinite Data

Our eyes in the sky – from Earth-watchers like Sentinel to deep-space explorers like the James Webb Space Telescope (JWST) and the future Square Kilometre Array (SKA) – are sending back tsunamis of data. JWST alone beams down enough information daily to fill a stack of laptops! Finding the faintest alien signal, spotting a subtle environmental shift, or classifying billions of stars in this data deluge is like searching for a single grain of sand on all the world's beaches. Classical algorithms are drowning.

Quantum Machine Learning (QML) offers a lifeline. Think of it as giving our analysis tools "quantum eyes," capable of spotting patterns and connections invisible to classical approaches. Imagine quantum algorithms sifting through petabytes of SETI data, identifying faint, structured signals buried deep within cosmic static – a potential whisper from across the void. Or picture QML analyzing satellite images of Earth with lightning speed, detecting the subtle fingerprints of deforestation, crop disease, or melting glaciers far faster and more accurately than before, helping us become better stewards of our own planet.

The Heart of the Matter: While the promise is huge, making QML truly outperform classical methods on messy, real-world data with today's noisy quantum machines is still a major research challenge. But the sheer volume drives the need: the SKA radio telescope network aims to generate more daily data traffic than the entire internet! Classical tools simply won't keep up.

Imagine the Scene: QML algorithms tirelessly scan astronomical surveys, automatically flagging unusual celestial objects or transient events for astronomers to investigate. In Earth science, they instantly correlate atmospheric data streams with ocean temperatures and land changes, dramatically improving weather forecasts and climate models, giving us a clearer crystal ball for our planet's future.

The quantum revolution in space won't happen overnight. Building these powerful "quantum brains" is one of the great scientific and engineering challenges of our time. But the journey has begun. From orchestrating the intricate dance of satellites to simulating the universe's birth and decoding cosmic whispers hidden in data floods, quantum computing promises to be an utterly transformative tool. It's about equipping ourselves with minds capable of not just calculating the cosmos, but truly understanding it, pushing the frontiers of exploration further than ever before. The final frontier is calling, and it speaks quantum.

Gravity's Secret Whispers: Listening to the Earth Below

We think of gravity as simple: apples fall down, we stay glued to our chairs. But that familiar pull isn't a flat, boring constant. It's a rich tapestry, subtly shifting depending on what lies hidden beneath our feet. Denser rock tugs a tiny bit harder; a hidden cave or a reservoir of water tugs a little less. For ages, detecting these minuscule variations meant using clever, clockwork-like devices – finely tuned springs or precisely dropped objects.

But now, we're entering the quantum age, and scientists, particularly across Europe, are building tools that listen to gravity's whispers with almost unbelievable clarity. Picture this: tiny clouds of atoms, like Rubidium, cooled by lasers to temperatures colder than deep space, reaching a state of near-perfect stillness. In this realm, atoms behave less like tiny balls and more like waves.

Using laser pulses like strokes of a microscopic choreographer, scientists guide these atom-waves along different paths and then bring them back together. Gravity's local strength affects their journey, leaving a subtle imprint – a "phase shift" – like ripples interfering on a pond. By reading these interference patterns, researchers can measure gravity with breathtaking precision. Labs like SYRTE in Paris are crafting portable versions of these "quantum gravimeters," sensitive enough to feel the gravitational difference of mere millionths of a percent!

Imagine the possibilities! These aren't just lab toys. Teams, like those at the University of Birmingham, are building rugged versions designed to leave the lab and explore the real world. They've already "seen" hidden tunnels underground in field tests. Think of it: mapping hidden water sources for thirsty communities, listening for the subtle swell of magma warning of a volcanic eruption, detecting dangerous sinkholes before roads collapse, or even spotting density changes that hint at buried mineral wealth. It's like giving humanity X-ray vision for the ground beneath us. Projects funded by the EU's Quantum Flagship and companies spun out of this research are pushing these incredible tools closer to everyday use, while grand projects like MIGA aim to build colossal underground atom antennas to eavesdrop on cosmic gravitational waves.

Finding Your Way When the Stars Go Dark: The Quest for Unshakeable Navigation

We rely on GPS like a digital compass pointing us home. But what happens when that signal is blocked, jammed, or simply unavailable – deep underground, underwater, or in a city's concrete canyons? Our digital lifeline goes dead. Recognizing this modern Achilles' heel, especially for defense, the US agency DARPA is on a bold quest: to create ways to navigate perfectly, anywhere, anytime, without needing an external signal like GPS.

Their dream is self-contained navigation. Think of the old inertial systems – gyroscopes and accelerometers tracking movement. They work, but tiny errors creep in, like a flawed memory, causing your calculated position to drift further and further off course over time. DARPA wants to banish this drift.

How? By turbocharging the components. They're funding the creation of atomic clocks – the ultimate timekeepers –

shrunk down onto chips, providing incredibly stable "heartbeats" for navigation calculations. They're even pushing towards optical clocks, potentially a thousand times more stable. Another clever trick is teaching systems to listen to the soup of signals already around us – TV, radio, satellites not meant for navigation – and triangulate position from this "noise."

And crucially, the quantum magic reappears. Those same atom interferometers sensing gravity can be tuned to be hyper-sensitive detectors of acceleration and rotation. Imagine an inertial system using quantum sensors, whose errors barely grow – perhaps only drifting meters in an hour, instead of miles. It's like giving a ship an internal compass and sextant so perfect it barely needs to see the stars. While the initial drive is defense, imagine autonomous cars that never get lost, emergency crews navigating disaster zones reliably, or critical systems staying perfectly synchronized even if external signals fail.

Giving Worlds an Ultrasound: Mapping Hidden Celestial Landscapes

Perhaps the most awe-inspiring use of gravity sensing takes us off-world. How do you map the inside of a planet or moon you can't drill into? You watch how it pulls on things. By precisely tracking spacecraft orbiting celestial bodies, scientists can detect those subtle gravitational variations we discussed earlier, but on a planetary scale.

NASA's GRAIL mission was a masterpiece of this. Two spacecraft flew in exquisite formation around the Moon, measuring the minute changes in the distance between them caused by mountains, craters, and denser areas below. The result? The most detailed gravity map ever created for any world, revealing a Moon far more complex than we knew – a shattered crust, hidden lava tubes, dense masses lurking beneath basins, and even clues about its tiny core. It was like giving the Moon an incredibly detailed internal scan, resolving features just miles across.

Missions to Mars have done similar work, mapping its crustal thickness, the immense gravity "bulge" of its giant volcanoes, the weight of its ice caps, and hints of where water might once have dramatically shaped its interior. Future explorers heading to mysterious asteroids or the potentially ocean-bearing moons of Jupiter and Saturn will undoubtedly use even more advanced gravity techniques – perhaps even deploying quantum gravimeters – to probe their structures, hunt for hidden oceans, and unlock the secrets of their birth.

The Emerging Sixth Sense

From the Earth's hidden depths to the heart of distant moons, and through the challenge of navigating our own world without familiar beacons, we're developing a new kind of perception. Driven by quantum physics and relentless innovation, we are learning to sense the unseen, to read the invisible forces that shape our universe. It's a journey that's expanding not just our knowledge, but our fundamental ability to interact with and understand the cosmos. We are tuning into gravity's whispers and the universe's subtle hum, opening up frontiers we are only beginning to imagine.

Gazing into the Quantum Void: Our Next Cosmic Frontier

Imagine the universe, not just as endless space dotted with stars, but as the ultimate, untamed laboratory. It's a place where the strangest rules of reality – the whispers of the quantum realm – might finally be coaxed into revealing their deepest secrets. Pushing our most advanced quantum technologies beyond the cradle of Earth isn't just an engineering challenge; it's the dawn of a new era of discovery. But this quantum space horizon, shimmering with revolutionary promise, presents hurdles that demand we rethink everything we thought we knew about building for the void.

Taming the Absolute Zero: The Deep Freeze Dilemma

Think of quantum states – the bizarre superposition and entanglement that power these future technologies – as the most delicate, fleeting phenomena imaginable. They're shy, often only appearing when temperatures plummet to near absolute silence, a mere fraction of a degree above the theoretical coldest possible point (-273.15°C or 0 Kelvin). Achieving this cosmic chill, often dipping into the milliKelvin range (thousandths of a degree above zero!), is tough enough in our ground-based labs. In space? It's a monumental feat.

Space offers a cruel paradox: while the background hums at a frigid 3 Kelvin, the vacuum makes shedding heat incredibly difficult. No air, no convection – just the slow bleed of radiation. Add to that the brutal constraints of spacecraft design – every gram, every watt of power is precious. The bulky, power-guzzling refrigerators common on Earth are simply too demanding for most cosmic voyages.

But human ingenuity thrives on constraint. Engineers are dreaming up new breeds of compact, ultra-efficient cryocoolers – think sophisticated descendants of the tech cooling instruments on the James Webb Space Telescope (JWST), but needing to reach temperatures thousands of times colder. They're layering insulation like cosmic onions, designing clever thermal switches, and pioneering multi-stage cooling systems that sip power, often running complex refrigeration on less energy than a few household lightbulbs.

And there's another trick up our sleeve: laser cooling. For clouds of atoms destined to become ultra-precise clocks or sensors, lasers can trap and chill them to microKelvin or even nanoKelvin temperatures, potentially sidestepping some of the bulkiest refrigeration needs. NASA's Cold Atom Lab (CAL) aboard the International Space Station is our stunning proof-of-concept. It's not just cooling atoms; it's creating Bose-Einstein condensates – an exotic state of matter where atoms act as one giant quantum wave – in the near-weightlessness of orbit, reaching effective temperatures colder than anywhere else in the known universe (down to picoKelvin!). Conquering this deep freeze is the key – the essential first step to unlocking quantum's power among the stars.

New Eyes on the Cosmos: The Dream of Quantum Telescopes

Our grandest telescopes, like Hubble and JWST, are magnificent, but they obey classical rules. Their sharpness is limited by the size of their mirrors and the light they gather. But what if we could sidestep those limits using the universe's own quantum weirdness? Enter the concept, still largely theoretical but utterly captivating, of the "quantum telescope."

Imagine not just collecting light, but harnessing the "spooky action at a distance" of entangled photons. Picture pairs of light particles, linked quantumly, captured by detectors flung vast distances apart across space. By measuring the correlations between these twins, theorists believe we could achieve an angular resolution – the ability to see fine detail – dictated by the separation between the detectors, not their individual size. This could mean seeing details hundreds or thousands of times finer than possible today. We could potentially map the surfaces of distant exoplanets or resolve the swirling chaos right at the edge of a black hole.

Related ideas like "quantum ghost imaging" add to the intrigue, offering ways to "see" an object without directly capturing the light that bounced off it, using entangled partners as proxies. Building such instruments is a staggering challenge – creating stable entangled photons, beaming them across space without breaking their fragile link, synchronizing measurements with impossible precision. Yet, the sheer potential, rooted in the fundamental laws of quantum mechanics, fuels the dream of giving humanity entirely new eyes on the cosmos.

From Bold Ideas to Orbiting Reality: Quantum Pioneers Take Flight

This isn't just science fiction chalk-talk; the journey has already begun. We're launching pathfinder missions, turning theory into tangible hardware orbiting above us.

Micius (China): Launched in 2016, this satellite became a quantum superstar. It proved we could beam entangled photons between space and Earth over 1200 kilometers, creating ultra-secure communication channels protected by quantum laws (Quantum Key Distribution - QKD). It acted as a relay, linking ground stations with quantum entanglement, and even successfully "teleported" quantum information from Earth to orbit. Micius didn't just work; it showed that delicate quantum states can survive the brutal trip through our atmosphere and the void, paving the way for a future global quantum internet.

Cold Atom Lab (CAL) (NASA): Since 2018, this physics lab on the ISS has been a quiet revolution. By leveraging microgravity, CAL allows scientists to observe ultra-cold atoms and Bose-Einstein Condensates for far longer durations (seconds instead of fractions of a second) than on Earth. It's a unique testbed, pushing the frontiers of cold matter physics to temperatures barely conceivable, and demonstrating the core technologies needed for future space-based quantum sensors – think ultra-precise navigation systems that don't rely on GPS, instruments to map Earth's gravity field with exquisite detail, or even probes to test Einstein's theories in new ways.

These missions are just the first explorers stepping onto the shores of the quantum space frontier. Future expeditions might carry quantum computers into orbit, deploy networks of entangled sensors to monitor our planet or listen for gravitational waves, or place atomic clocks in space whose precision could redefine timekeeping and fundamental physics.

The quantum space horizon is vast, challenging, and electrifying. Overcoming the deep freeze, inventing new ways to see, and learning from each pioneering flight, we are slowly but surely equipping ourselves to explore the universe not just as observers, but as participants in its deepest quantum dances. The discoveries waiting in that cold, quantum-tinged expanse could rewrite our cosmic story.

About Author

Azhar ul Haque Sario is a bestselling author and data scientist with a remarkable record of achievement. This Cambridge alumnus brings a wealth of knowledge to his work, holding an MBA, ACCA (Knowledge Level - FTMS College Malaysia), BBA, and several Google certifications, including specializations in Google Data Analytics, Google Digital Marketing & E-commerce, and Google Project Management.

With ten years of business experience, Azhar combines practical expertise with his impressive academic background to craft insightful books. His prolific writing has resulted in an astounding 2810 published titles, earning him the record for the maximum Kindle editions and paperback books published by an individual author in one year, awarded by Asia Books of Records in 2024.

ORCID: https://orcid.org/0009-0004-8629-830X
Azhar.sario@hotmail.co.uk
https://www.linkedin.com/in/azharulhaquesario/

www.ingramcontent.com/pod-product-compliance
Lightning Source LLC
Chambersburg PA
CBHW031345230926
51848CB00002B/390
9783384582638